T0335754

CLIMATE CHANGE AND THE ENERGY PROBLEM

Physical Science and Economics Perspective

Second Edition

CLIMATE CHANGE AND THE ENERGY PROBLEM

Physical Science and Economics Perspective

Second Edition

David Goodstein
California Institute of Technology, USA

Michael Intriligator

World Scientific

NEW JERSEY · LONDON · SINGAPORE · BEIJING · SHANGHAI · HONG KONG · TAIPEI · CHENNAI · TOKYO

Published by

World Scientific Publishing Co. Pte. Ltd.

5 Toh Tuck Link, Singapore 596224

USA office: 27 Warren Street, Suite 401-402, Hackensack, NJ 07601

UK office: 57 Shelton Street, Covent Garden, London WC2H 9HE

Library of Congress Cataloging-in-Publication Data
Names: Goodstein, David L., 1939–   author. | Intriligator, Michael D., author.
Title: Climate change and the energy problem : physical science and economics perspective /
    by David Goodstein (Caltech), Michael Intriligator (University of California, Los Angeles, USA).
Description: 2nd Edition. | New Jersey : World Scientific, [2017] |
    Revised edition of the authors' Climate change and the energy problem, c2013.
Identifiers: LCCN 2016056676| ISBN 9789813208346 (hc : alk. paper) |
    ISBN 9789813209800 (pbk : alk. paper)
Subjects: LCSH: Power resources. | Energy industries. | Climatic changes. |
    Global warming. | Renewable energy sources.
Classification: LCC HD9502 .G66 2017 | DDC 333.79--dc23
LC record available at https://lccn.loc.gov/2016056676

British Library Cataloguing-in-Publication Data
A catalogue record for this book is available from the British Library.

Desk Editor: Kalpana Bharanikumar

Typeset by Stallion Press
Email: enquiries@stallionpress.com

Printed in Singapore

Contents

Introduction

The first edition of this book made some assumptions that have not quite stood the test of time. Here we'll take a fresh look at those suppositions.

As noted in the book, oil extraction in the United States had reached a peak around 1970 and decreased rapidly after that. In the last few years, however, the rate of oil extraction in the U.S. took off once again and continues to rise toward a new peak. The reasons include the perilous business of deep-water drilling and the large-scale and in many respects equally perilous, application of hydraulic fracturing.

A large drop in the price of oil took place in the year 2014. Countries like Russia, Venezuela and Iran that had counted on the price of oil reliably remaining at $100 or more per barrel found their economies thrown into turmoil by this unexpected and unwelcome development. The primary cause of the fall was that oil extraction within the United States had increased to the point where it once again reached the level of the original Hubbert's peak. The result was a huge disruption in the cost and availability of oil everywhere.

The big increase in U.S. extraction is largely due to the widespread use of hydraulic fracturing, colloquially known as fracking, to obtain oil and gas that was not previously accessible. Fracking consists of the use of water, sand and chemicals under high pressure to break up and extract otherwise unattainable deposits of oil and gas. Its hefty consumption of water is especially worrisome, considering that fresh water is in increasingly short supply everywhere. Along with its use of vast quantities of water, one of the most critical issues associated with fracking is its impact on groundwater contamination. Wastewater disposal is one of its major problems: Something like 10–40% of the chemical mixture injected into the ground during fracking flows back to the surface during well development although further research is needed to conclusively determine fracking's role in groundwater contamination, as well as its impact on climate change and air pollution, it is already clear that fracking, its ability to generate oil notwithstanding, presents all of us with major problems.

In particular, people living near fractured wells are potentially at risk of health threats from the increased amount of volatile organic compounds and air toxins that the fracking releases into the immediate environment. In fact, the use of fracking has been outlawed in a number of states starting with New York, and in many other locations. For example, the practice has been banned in Vermont and Maryland, as well as in a number of jurisdictions in California and Texas, to say nothing of Wales and Scotland. There is even evidence of fracking causing earthquakes in Oklahoma.

In addition to widespread fracking, the United States has also seen a large increase in the drilling of dangerous deep-water oil wells. There are approximately 3,400 such wells with depths greater than 150 m in the Gulf of Mexico alone. The sheer size and depth of these rigs are cause for concern.

As is well known, in April 2010, a BP (British Petroleum) deep-water oil rig exploded, killing 11 people and releasing 750,000 m^3 (200 million gallons) of oil into the Gulf of Mexico. The environmental impact was devastating and widespread, including a large number of animal deaths resulting from the release of the oil. It is estimated that over 82,000 birds, about 6,000 sea turtles and nearly 26,000 marine mammals were killed from either the initial explosion or the oil spill. Many scientists consider that event to be one of the worst environmental disasters in the history of the United States. Although notable for its catastrophic scale, the Gulf of Mexico spill was far from an isolated event. The U.S. Minerals Management Service calculates that deep-water drilling has been responsible for 600 oil spills, including 9 large ones, since 2007.

Before the turn of the century, deep-water wells were generally not considered to be either technologically or economically feasible, but as oil prices climbed over the last two decades, more energy companies started investing in this area. The deep-water gas and oil market is back on the rise after the 2010 Gulf of Mexico disaster, with total expenditures of around $35 billion per year and a total global investment of $167 billion in the past four years. It has yet to be seen whether fluctuations in the price of oil will call a halt to deep-water drilling.

For a long time there were no new nuclear energy plants in the United States, but recently construction has commenced on a small number of new facilities, although their numbers are nowhere near enough to substitute for the use of fossil fuels. There are a number of possible new sources of fossil fuels in addition to that gained by fracking and deep water drilling, including the oil sands of Alberta, Canada and so-called shale oil, which is not really oil at all, as well as some new discoveries of conventional oil sources, among them rather large deposits in North Dakota. Finally, the use of ethanol has somewhat extended our supply of fuels.

Nevertheless, this apparent bonanza is strictly temporary. Nature's supply of oil and gas remains finite, and eventually we humans will run out of new sources to develop. A new Hubbert's peak for fracking and deep-water drilling and the rest will follow the earlier Hubbert's peak for conventionally, more easily available oil. We can still expect that the total worldwide oil production as a function of time will be a bell-shaped curve, with a peak in the not too distant future. Nobody knows for sure when that peak will occur, but with China, India and other nations turning increasingly to cars and ramping up their consumption of fuel, the world-wide demand for oil will continue its runaway increase until the new peak is reached.

CHAPTER 1

THE SCIENCE OF ENERGY
AND GLOBAL WARMING

The world has started to run out of conventionally produced, cheap oil. Moreover, burning up all that oil has greatly increased the atmosphere's concentration of carbon dioxide, a greenhouse gas. The first of these problems has been faced by resorting to deep water drilling, the use of tar sands and hydraulic fracturing (fracking) all of which have resulted in a temporary increase in the amount of oil available. People are trying to solve the second problem by means of international treaties, which so far have had limited success.

As the 2010 oil spill disaster in the Gulf of Mexico showed vividly, oil drilling in deep water and other questionable locations is extremely dangerous. It took hundreds of millions of years for the Earth to build up the supply of oil that we started with and we have used up roughly half of the accessible supply in a mere 150 years. No one knows how long the less accessible supply will last, but even in the absence of drilling disasters, it will not last forever.

If we overcome the shock of the end of inexpensive oil and turn to coal and natural gas, which are the other two forms of fossil fuel, life may go on as it always has until we start to run out of those resources as well. And by the time we have used up all the fossil fuels the greenhouse effect we increase by then may well render the planet unfit for human life. Even if life does go on, civilization as we know it will not survive unless we can find a way to live without fossil fuels.

Technically, it might be possible to accomplish that. Power plants can run on nuclear energy and even after that is gone, there will always be sunlight and its derivatives, such as wind and hydroelectric power, sources that will not contribute at all to global warming. Part of that power can be used to generate hydrogen or charge advanced electric batteries for use in transportation. Those sources too, will not contribute to global warming.

There are huge technical problems to be solved for all of this to work, but most of the necessary scientific principles are well understood and we humans are very good at solving technical problems. In fact, if we put our minds to it, we could start trying to kick the fossil fuel habit now, protecting the planet's climate from further damage and preserving the fuels for future generations as the source for chemical goods. Some 90% of the organic chemicals we use, including pharmaceuticals, agricultural chemicals and plastics are made from petroleum. There are clearly better uses for the stuff than burning it up.

To make such an about-face will take global political leadership that is both visionary and courageous. It seems unlikely that we will be so lucky. A few years ago, no one was paying any attention to this problem. Now everybody pays lip service to it, but precious little is being done about it.

Oil is the most important of the fossil fuels. Its existence has been known about for thousands of years because it was found in natural seepages at the surface of the Earth. Ancient people in the Middle East and the Americas used oil for a variety of medical, military and other purposes. It was thought to be useful as a laxative for example (do not try that at home!). The Persians in the siege of Athens in 480 B.C.E. used oil soaked blazing arrows. But by and large, oil was little needed and little used until the nineteenth century.

By the beginning of the nineteenth century, the growth of urban centers made it necessary to search for a better means of illumination, whose forms had barely changed since antiquity. For a while, whale-oil lamps served the purpose and whaling became a significant industry. But by the middle of the century, whales were becoming scarce. Kerosene derived from coal was also widely used, but a better substitute for whale oil was needed. In August 1859, Edwin L. Drake, a former conductor on the New York and New Haven railroad, drilled the world's first successful oil well near Titusville in Northwestern Pennsylvania. Soon coal–oil refineries were processing cheap oil instead of coal; and oil became widely used for illumination as well as lubrication.

Then in 1861, German entrepreneur Nikolaus Otto invented the first gasoline-burning engine, the direct ancestor of the engines in our cars today; and soon demand for oil as a fuel began to grow. Within a few decades, oil was being found in and extracted from fields all over the globe. Since E. L. Drake drilled that first well, roughly 50,000 oil fields have been discovered worldwide, but most of the discoveries have been insignificant. About half the oil ever discovered has been found in the 40 largest fields.

In the 1950s, Shell Oil Company geophysicist Marion King Hubbert predicted that the rate at which oil could be extracted from wells in the United States would peak around 1970 and decline rapidly after that. At that time his prediction was not well received by his peers, and it was dismissed, but he turned out to be uncannily accurate. United States oil extraction peaked at around nine million barrels per day in 1970 and started to decline. But, as we have seen, new and dangerous methods of drilling for oil have been applied, with the result that we are headed for a new peak. All attempts to predict a worldwide peak will be subject to the same shortcoming, because the same methods of drilling are being applied everywhere.

Nevertheless, Hubbert's followers have succeeded in making a crucial point — the worldwide supply of oil, as with any mineral resource, will rise from zero to a peak and after that it will decline forever. We will be in trouble not when we pump the last drop, but when we reach roughly the halfway point and the amount we can extract begins to drop, while the insatiable demand for oil continues to rise.

Some say that the world has enough oil to last for another few hundred years or more, but that view may be mistaken. The peak, which will occur when we have used up roughly half of all the oil made by nature for us, may well come far sooner than that. When the peak occurs, increasing demand will meet decreasing supply, with disastrous results. We had a foretaste of the consequences in 1973 when some Middle Eastern nations took advantage of declining United States supplies and created a temporary, artificial shortage. The immediate result was long lines at gas stations and despair for the future of the American way of life. After the worldwide Hubbert peak, the shortage will not be artificial and it will surely not be temporary.

There are those who see a silver lining in this dire situation. Since the beginning of the Industrial Revolution, we have been pouring carbon dioxide and other greenhouse gases into the atmosphere precisely because we have been burning fossil fuels. This has resulted in an increase in global temperature that will continue and might accelerate. Could it be that Hubbert's peak will prevent us from destroying our planet?

The climate of the Earth is in a fragile, metastable condition that probably was created by life itself. Primitive life forms were responsible for oxygenating the atmosphere and they were also responsible for laying down huge quantities of carbon in the form of coal and other fossil fuels. If after Hubbert's peak we take to burning coal in large quantities, then Earth's so-called intelligent life will be reconverting that carbon and oxygen into carbon dioxide. We cannot predict exactly what that will do to our climate, but one possibility is that it will throw the planet into an entirely different state. The planet Venus is in such a state because of a runaway greenhouse effect that has rendered its surface temperature hot enough to melt lead. We have a grave responsibility to prevent the same thing from happening on Earth.

Some economists say that we do not need to worry about running out of oil because while it is happening, the rise in oil prices will make other fuel economically competitive and oil will be replaced by something else. However, they do not tell us what that something else might be. There is nothing on the horizon that looks like an adequate substitute for cheap oil. And besides, as we learned in 1973, the effects of an oil shortage can be immediate and drastic. Meanwhile, it may take many years, perhaps even decades, to replace the vast infrastructure that supports the manufacture, distribution and consumption of the products of the 20 million barrels of oil that Americans alone gobble up each day. One certain effect of the coming shortage of cheap oil will be steep inflation, because gasoline, together with everything made from petrochemicals and everything that has to be transported, will suddenly cost more. Such an inflationary episode will certainly cause severe economic damage, perhaps so severe that we will be unable to replace the world's vast oil infrastructure with something else. That is a prospect we would rather not think about.

Let us have a look at the sources of oil that have come on line recently. Canada, for example, has large supplies of tar sands in its province of Alberta, which have been renamed as oil sands to attract investment. These are solid deposits that must be strip-mined at enormous cost to the environment. Even then, it takes two tons of ore to make one gallon of a substance that is not rich enough to distill into gasoline, so hydrogen must be added. Accordingly, Alberta has some of the largest plants in the world for extracting hydrogen from methane. The net result is that even though Canada has copious supplies of this substance, it will not easily replace the cheap oil once that resource starts to run dry.

Oil found deep beneath the ocean presents special problems and dangers, as the 2010 oil spill in the Gulf of Mexico made clear. Nuclear power plants are so expensive, feared and controversial that none has been built in the United States for many years and some countries (for example, Italy) have outlawed them completely. The recent disaster at the Fukushima Daiichi power plant in Japan has made things even worse. When the oil crisis comes, opposition to nuclear power is likely to weaken considerably, but it will take at least a decade or more for the first new power plants to come online (something like 20 new nuclear power plants are currently under construction in the United States).

CHAPTER 2

ENERGY MARKET STRUCTURES
AND REGULATORY STRUCTURES

Today, energy is universally recognized as an essential resource in every modern economy such as those of Western Europe, North America and Australasia, and in addition, the modernizing economies of China and India. As noted in the Chapter 1, changes in the availability and price of energy have major impacts on the environment, on standards of living and economic growth and indeed, on every aspect of economies that rely on accessible energy supplies to sustain and to improve the quality of life of their citizens.

Before the shock of the 1973 Organization of Petroleum Exporting Countries (OPEC) oil embargo, discussed in greater detail below, the extent to which energy problems could disrupt the comforts of modern life was not readily appreciated. In the United States, for example, energy was generally obtainable at a relatively low cost and its easy availability was taken for granted. Some warned of problems ahead, but there was little immediate concern and those who did sound alarms were widely dismissed as Cassandras and shunned by the oil industry as well as governments. All that changed with the OPEC embargo, which caused oil prices to quadruple between 1973 and 1975. The 1979 revolution in Iran, which overthrew America's longtime ally, the Shah, once again led to major disruptions in oil supplies and an even more pronounced price spike. The 1973 embargo raised oil prices by about $5.50 a barrel, from about $4.00 to over $9.50. At that time, Americans consumed about 17.3 million barrels of oil a day, so that the increase amounted to an oil levy of $35 billion, or 2.5% of gross national product (GNP).

The 1979 price shock was even bigger, with oil prices rising by about $21 a barrel, equivalent to a levy of $144 billion on United States users, or about 6.5% of GNP. These events finally forced governments and citizens throughout America and the Western world to recognize the problems

7

associated with energy scarcity and to acknowledge the damaging potential
of future energy crises. Meanwhile, in the United States, oil production for
the lower 48 states had reached the peak predicted by Hubbert in 1970.
It would recover later as we have seen, but nevertheless, within a few years,
the public would also become far more aware of the relationship between
energy policies and environmental concerns, which constitute the main focus
of this book.

Almost all energy products have a market structure characterized by a
relatively small number of intensely competitive major producers, a larger
number of smaller producers and an abundance of consumers. Economists
refer to this type of market structure — one with few sellers and many
buyers — as an oligopoly. Within the oil industry, several major producers
dominate the market for refining, shipping and distribution. They include
the multinational oil companies — the so-called "supermajors" of BP
(formerly British Petroleum), Chevron, ExxonMobil, ConocoPhillips, Shell,
Eni and Total S.A., as well as many so-called "majors" and "independents".
They are all highly competitive with one another in all aspects of the industry, vying for everything from drilling rights to marketing opportunities.

As to oil reserves, the largest of these are nearly all controlled today
by national oil companies, including, in descending order of the size of
their reserves, the Saudi Arabian Oil Company, the National Iranian Oil
Company, the Qatar General Petroleum Corporation, the Iraq National Oil
Company, Petroleos de Venezuela S.A., the Abu Dhabi National Oil Company, the Kuwait Petroleum Corporation, the Nigerian National Petroleum
Corp., the National Oil Company (Libya) and Sonatrach (Algeria). It goes
without saying that these large oil-producing national entities form an
oligopoly for the world crude oil market. Anytime they collectively agree
to reduce their supply of crude, the price of oil rises.

In this oligopoly situation, with few producers, each firm can have an
influence on price. For example, if Saudi Arabia, which is a major producer
of crude oil, decides to reduce its production that has the potential to
significantly affect the world price of oil. Ultimately, however, the impact
on price would depend on what other producers do in response. If they
increase their production to offset the Saudi cuts, oil prices would not be
significantly affected. Thus, all the major producers must consider what
their fellow producers would do, a situation referred to by economists as
"mutual dependence recognized". Such recognition requires all the firms in
the industry to take into account the probable or even possible behavior and
reactions of the others, a situation of strategic behavior. Economists study

such behavior using the tools of game theory, looking at the various players' possible strategies and at how the other players, or actors, might counter them. Some or all of the actors might collude to deal with this situation: The OPEC cartel is an example of such collusion.

The major oil companies are motivated by profit from the sale of their energy and as a result, are focused on their market share. They know from experience that when exploration yields insufficient reserves to maintain their share of the market, they must start to push for alternative sources of supply, such as ethanol. These would also include energy sources typically financed by governments, such as nuclear power and more recently, emerging wind and solar power technologies.

As to the buyers of final products of the energy industry, whether for oil, gas, or coal, they are a mixture of very large firms that also represent a type of oligopoly and small entities that tend to be closer to the perfectly competitive model of firms that do not collude. Thus, there is a strong asymmetry between the sellers and the buyers, with the sellers having much greater influence than the buyers.

The structure of the natural gas market is also one of oligopoly, with a handful of major national suppliers: Russia, the United States, the Arab League, the European Union, Canada, Iran, Norway, Algeria, Qatar, the Netherlands, Saudi Arabia and Indonesia. The product is shipped via pipeline or in the form of Liquefied Natural Gas (LNG) and many competing firms distribute it to consumers. It is for the most part consumed in the countries in which it is produced or sold on an inter-regional basis within contiguous countries or trading blocs. For example, Russian gas is shipped by pipeline to Eastern and Western Europe. Gazprom, by far the largest firm in Russia and one of the largest energy-supplying firms in the world, controls this supply. It has in recent years started to supply oil as well as natural gas.

In the United States, the energy industries, including oil, natural gas and coal, are regulated at the local, state and national level. Locally, permits are issued to those who seek to explore for energy products. The standards for regulation at the statewide level are based on those originally established in the 1930s for the Texas oil fields by the Railroad Commission of Texas, which, despite its name, has played the leading role in regulating the state's production of oil and natural gas. Historically, these standards favored domestic producers by ensuring that imported oil prices remained higher than those of domestic oil, requiring consumers to pay more for externally produced oil.

In 1890, the United States Congress passed the Sherman Antitrust Act to regulate monopolies, specifically the petroleum industry. Its actions were aimed primarily at Standard Oil, which, under John D. Rockefeller, had initially established a monopoly on the refining of oil and subsequently expanded into every aspect of the petroleum industry, from exploring to retailing. President Theodore Roosevelt first moved against Standard Oil as a monopoly in 1901 and in a 1911 landmark decision, the Supreme Court found that the company's policies had violated antitrust legislation and ordered that the firm be broken up. This decision eventually had impacts on price manipulation as the successor companies started to compete. Many of these companies still exist as Exxon (formerly Standard Oil of New Jersey), Mobil (formerly Standard Oil of New York), Chevron (formerly Standard Oil of California) and Amoco (formerly Standard Oil of Indiana). In 1914, the Clayton Antitrust Act was passed to prevent monopolies from forming through mergers. Today, the Antitrust Division of the Department of Justice and the Federal Trade Commission enforce this antitrust legislation and the earlier Sherman Act. The late 1970s and particularly the years of the Reagan administration saw an increasing trend toward widespread deregulation. Crude oil prices and refined petroleum products were deregulated in 1981.

Internationally, this was also a period of considerable change and ferment on the energy front. The 1970s witnessed the rise of OPEC, a cartel of oil-producing nations in the Middle East, Asia, Africa and Latin America. OPEC was actually founded in the year 1960 with the assistance of the United States, but gained control of crude oil supplies in the early 1970s through its member nations' takeovers of the major private oil companies that had established its oil fields. It continues to be extremely influential in the international oil industry, coordinating its members' crude oil production activities and attempting to limit production to ensure that prices make exports profitable. Collective action on the part of its members can cause a surplus or shortage of oil on world markets, which in turn affects oil prices.

The OPEC cartel first struck the West as a force to be reckoned with after the 1973 Arab–Israeli war, when its seven Arab nations, meeting that October in Kuwait, announced a 5% cutback in supply, a sharp price increase and an embargo on countries that had supported Israel. A second embargo, touched off by the Iranian revolution in 1979–1980, once again created artificial oil shortages, leading to long lines for higher priced gas. Yet another oil price shock occurred in August 1990, when Iraq invaded

Kuwait, and the supply of Saudi oil was threatened. The price of oil rose sharply that year, from $21 a barrel at the end of July 1990 to a peak of $46 in mid-October. Traders and the monetary authorities correctly viewed that rise as temporary. Since then, the power of OPEC has been curtailed by the emergence of new major suppliers of oil who are not members, including the United Kingdom, Mexico, Norway, Canada and Russia. The United States economy relied heavily on imported oil, but most of it came from the non-OPEC suppliers: Canada, Mexico, the UK and Russia, although OPEC members Nigeria, Saudi Arabia and Venezuela were also suppliers. More recently the U.S. oil supplies have greatly increased as a result of extraordinary drilling techniques. Furthermore, as noted earlier, OPEC members do not always act in concert, and defections from the cartel by one or more members, can cause cartel agreements to unravel, leaving oil supplies and prices largely unaffected.

Nevertheless, OPEC nations in the Persian Gulf area continue to hold the world's key reserves and to exert great influence on oil production and prices. The United States and many other nations that relied more and more on imported oil were increasingly vulnerable to such disturbances, as sharply higher prices brought about by the Spring 2011 upheavals across the Middle East and North Africa duly demonstrated. Recognizing this vulnerability after the 1973–1974 oil embargo, the United States established the Strategic Petroleum Reserve (SPR) in 1975 to mitigate future temporary supply disruptions. The United States at one time imported a net 12 million barrels $(1,900,000\,\mathrm{m^3})$ of oil a day so the SPR held about a 58-day supply. However, the maximum total withdrawal capability from the SPR was only 4.4 million barrels $(700,000\,\mathrm{m^3})$ per day, making it a 160+ day supply. With the great increase in drilling recently, that is, for the time being, no longer a problem. With a capacity of 727 million barrels, the federally owned oil stocks, which constitute the world's largest supply of emergency crude, are stored in huge underground salt caverns along the coastline of the Gulf of Mexico.

Internationally, the Paris-based International Energy Agency (IEA) implements and oversees a program of international energy cooperation among its 28 member nations, which include many members of the European Union, as well as the United States, Canada, Australia, New Zealand, Japan, South Korea and Turkey. It provides estimates of world oil demand and monitors changes in supply and demand over time. Its member governments take joint measures to meet oil-supply emergencies, as well as share energy information, coordinate energy policies and cooperate in the development of programs designed to promote energy security, to encourage economic growth and to protect the environment. IEA nations are also committed to improving systems for coping with oil supply disruptions, to promoting rational energy policies through cooperative relations with non-member countries, industry and international organizations and to operating a permanent information system on the international oil market. They also work to improve the world's energy supply-and-demand structure by developing alternative energy sources, increasing the efficiency of energy use, promoting international collaboration on energy technology, and assisting in the integration of environmental and energy policies.

CHAPTER 3

THE FUTURE

In the 1950s, the United States was the world's leading producer of oil. Much of the nation's industrial and military might derived from its giant oil industry. The country seemed to be floating on a rich, gooey ocean of "black gold". Nobody was willing to believe that the party would ever end. Well, almost nobody. There was a geophysicist named Marion King Hubbert who knew better.

Hubbert, the son of a central Texas farm family, was born in 1903. Somehow, he wound up at the far off University of Chicago, where he earned all his academic degrees right up to the Ph.D. Embittered with the academic world after an unhappy stint teaching geophysics at Columbia University, he spent the bulk of his career with the Shell Oil Company of Houston. That is where he was when, in 1956, very much against the will of his employer, he made public his calculation that American oil dominance would soon come to an end. He predicted that the rate at which oil could be extracted from American wells would reach a peak (now called "Hubbert's Peak") around 1970, and would decline rapidly after that. He turned out to be precisely correct, although as we have seen U.S. oil production has picked up again and is on its way to a new peak.

To understand how he reached that conclusion and the relevance of his reasoning to world oil supplies today, we need to understand a bit about how oil came to be in the first place. For hundreds of millions of years, animal, vegetable and mineral matter drifted downward through the waters to settle on the floors of ancient seas. In a few privileged places on Earth, strata of porous rock formed that were particularly rich in organic inclusions. With time, these strata were buried deep beneath the seabed. The interior of the Earth is hot, heated by the decay of natural radioactive elements. If the porous source rock sank just deep enough, it reached the proper temperature for the organic matter to be transformed into oil. Then the weight of the rock above it could squeeze the oil out of the source rock like water out of a sponge, into layers above and below where it could be trapped. Over vast stretches of time, in various parts of the globe, the seas retreated, leaving some of those deposits beneath the surface of the land.

Oil consists of long chain hydrocarbon molecules. If the source rock sank too deep the excessive heat at greater depths — some three miles below the surface — broke these long molecules into shorter molecules, which we call natural gas. Meanwhile, in certain swampy places, the decay of dead plant matter created peat bogs. In the course of eons, buried under sediments and heated by the Earth's interior, the peat was transformed into coal, a substance that consists mostly of elemental carbon. Coal, oil and natural gas are the primary fossil fuels. They are energy from the Sun, stored within the Earth.

Until only 200 years ago — the blink of an eye on the scale of our history — the human race was able to live almost entirely on light as it arrived from the Sun. The Sun nourished plants, which provided food and warmth for us and our animals. It illuminated the day and (in most places) left the night sky sparkling with stars, to comfort us in our repose. Back then, a few people in the civilized world traveled widely, even sailing across oceans, but most people probably never strayed very far from their birthplaces. For the rich, there were beautiful paintings, sophisticated orchestral music, elegant fabrics and gleaming porcelain. For the common folk, there were more homespun versions of art, music, textiles and pottery. Merchant ships ventured to sea carrying expensive and exotic cargoes, including spices, slaves and in summer, ice. At the end of the eighteenth century, no more than a few 100 million people populated the planet. A bit of coal was burned, especially since trees had started becoming scarce in Europe (they soon would begin to disappear in the new world as well) and small amounts of oil that seeped to the surface found some application, but by and large, Earth's legacy of fossil fuels was left untouched.

Today, we who live in the developed world expect illumination at night and air conditioning in summer. We may work every day up to 100 miles from where we live, depending on multiton individual vehicles to transport us back and forth, on roads paved with asphalt (another by-product of the age of oil). Thousands of airline flights a day take us to any destination on Earth in a matter of hours. When we get there we can still chat with our friends and family back home, or conduct business as if we had never left the office. Amenities that were once reserved for the rich are now available to most people. Refrigeration rather than spices preserves food and machines do much of our hard labor. Ships, planes, trains and trucks transport goods of every description all around the world. We do not see the stars so clearly anymore, but on most counts, few of us would choose to return to the eighteenth century.

This revolutionary change in our standard of living did not come about by design. If you asked an eighteenth century sage like Benjamin Franklin what the world really needed, he probably would not have described those things we wound up with — except perhaps for the dramatic improvement in public health that has also taken place since then. Instead of design or desire, our present standard of living arose from a series of inventions and discoveries that altered our expectations. What we got was not what we wanted or needed but rather what nature and human ingenuity made possible for us. One consequence of those inventions and changed expectations is that we no longer live on light as it arrives from the Sun. Instead, we are using up the fuels made from sunlight that Earth stored up for us over those many hundreds of millions of years. Obviously, we have unintentionally created a trap for ourselves. We will, so to speak, run out of gas. There is no question about that. There is only a finite amount of oil left in the tank. When will it happen? No one really knows. The supply of easily available oil has already started to run out and we are increasingly dependent on more difficult sources.

Throughout the twentieth century, the demand for and supply of oil grew rapidly. Those two are essentially equal. Oil has always been used as fast as it is pulled out of the ground. Until the 1950s, the oil supply continued to increase and oil geologists maintained the mathematically impossible expectation that the same rate of increase could continue forever. All warnings of finite supplies were hooted down because new reserves were being discovered faster than consumption was rising. Then, in 1956, Hubbert predicted that the rate at which oil could be easily be extracted from the lower 48 United States would peak around 1970 and decline rapidly after that. When, against all expectations, his prediction turned out to be right on, other geologists started paying serious attention.

Hubbert had used a number of methods to do his calculations. The first was similar to ideas that had been used by population biologists for well over a century. When a new population — of humans or any other species — starts growing in an area that has abundant resources, the growth is initially exponential, which means that the rate of growth increases by the same fraction each year, like compound interest in a banking account. That is just how geologists used to think oil discovery would continue to grow.

However, once the population is big enough so the resources no longer seem unlimited, the rate of growth starts slowing down. The same happens with oil discovery because the chances of finding new oil get smaller when there is less new oil to find. Hubbert showed that once the increase of oil discovery starts to decline, it is possible to extrapolate the declining rate to find where growth will stop altogether. At that point all the easily discoverable oil in the ground has been discovered and the total amount there ever was is equal to the amount that is already been used plus the known reserves still in the ground. Hubbert noticed that the trend of declining annual rate of increase of oil discovery was established for the lower 48 states by the 1950s. Others have now pointed out that the total quantity of conventional oil discovered worldwide has been declining for decades. The total quantity of conventional oil that the Earth stored up for us is estimated by this method to be about two trillion barrels. Of course, unconventional sources of oil may extend that, but nobody knows by how much.

Hubbert's second method required assuming that in the long run, when the historical record of the rate at which oil was pumped out of the ground was plotted, it would be a bell shaped curve. That is, it would rise at first and then reach a peak that would never be exceeded. Of course now it has been exceeded, but that's another story.

Hubbert's third method applied the observation that the total amount of oil extracted to date paralleled oil discovery, but lagged behind by a few decades. In other words, we pump oil out of the ground at the same rate at which we discover it, but a few decades later. Thus, the rate of discovery predicts the rate of extraction. Worldwide, remember the rate of discovery started declining decades ago. In other words, Hubbert's peak for oil discovery already occurred, decades ago. The new peak for the discovery of more difficult sources of oil is not yet obvious.

Not all geologists agreed with this assessment. Many preferred to take the total amount known for sure to be in the ground and divide that by the rate at which it is getting used up. This is known in the industry as the R/P ratio — that is, the ratio of reserves to production. Depending on what data one uses, the R/P ratio is currently between 40 and 100 years. They conclude that if we continue to pump oil out of the ground and consume it at the same rate we are doing now, we will not have pumped the last drop for another 40–100 years. That may well be true, but running out of all oil in 40–100 years is not a pretty prospect.

The fact is the amount of known reserves is a very soft number. For one thing, it is usually a compilation of government or commercial figures from countries around the world and those reported figures are at least sometimes distorted by political or economic considerations. For example, the Royal Dutch Shell Corporation was recently deeply embarrassed when it was forced to revise downwards its "known reserves". That can happen to a corporation that is subject to audits by national bodies, but nobody ever audits the figures produced by nations themselves.

Nevertheless all our experience with the consumption of natural resources suggests that the rate at which we use them up starts at zero, rises to a peak that will never be exceeded, and then declines back to zero as the supply becomes exhausted. That picture forms the fundamental basis of the views of Hubbert and his followers, but those who depend on the R/P ratio ignore it. Given that the worldwide demand will continue to increase as it has for well over a century, especially with China and India coming on line as car-driving nations, Hubbert's followers expect the crisis to occur when the peak is reached, rather than when the last drop is pumped. In other words, we will be in trouble when we have used up half the oil that existed, not all of it.

If you believe the Hubbert's view — that the crisis comes when we reach the production peak rather than the last drop, then if Hubbert's followers are correct, we may be in for some difficult times in the future. In an orderly, rational world, it might be possible for the gradually increasing gap between supply and demand for oil to be filled by some substitute, like sunlight, wind- or other renewable power. But anyone who remembers the oil crisis of 1973 knows that we do not live in such a world, especially when it comes to an irreversible shortage of oil. It is impossible to predict exactly what will happen, but we can all too easily envision a dying civilization, with its landscape littered with the rusting hulks of all kinds of gasoline driven cars. Worse, desperate attempts by one country or region to maintain its standard of living at the expense of others could lead to Oil War III (Oil Wars I and II are already history). Knowledge of science is not useful for predicting whether or not such dire events will occur. Science is useful, however, in placing limits on what is possible.

We are already using in our cars a fuel that may require more energy to produce than it provides. Ethanol, made from corn, is widely believed to be a net energy loser. As we proceed down the fossil fuel list from light crude oil (the stuff we mainly use now) to heavy oil, oil sands, tar sands and finally shale oil, the cost in energy progressively increases, as do other costs. (**Shale oil** is an unconventional oil produced from oil shale rock fragments by converting the organic matter within the rock (kerogen) into synthetic oil and gas). Some experts believe that shale oil will always be an energy loser.

Once past Hubbert's final peak, as the gap between rising demand and falling supply grows, the rising price of oil may make those alternative fuels economically competitive, but even if they are net energy positive it may not prove possible to get them into production fast enough to fill the growing gap. That is called the "rate-of-conversion problem". Worse, the economic damage done by the rapidly rising oil prices may undermine our ability to mount the huge industrial effort needed to get the new fuels into action.

Natural gas, which comes from overcooked source rock, is another alternative in the short term. Natural gas, mostly methane, is relatively easy to extract quickly, and transformation to a natural gas economy could probably be accomplished more easily than is the case for other alternative fuels. Ordinary engines, similar to the ones used in our cars can run on compressed natural gas. Alternatively, natural gas can be converted chemically into a liquid that could substitute for gasoline (liquid natural gas is a low temperature liquid that requires refrigeration and special handling. We are referring here to a liquid made chemically from natural gas). Even so, replacing the existing vehicles and gasoline distribution system, or building the industrial plant to convert methane to gasoline fast enough to make up for the missing oil will be difficult. Even if this transformation is accomplished, success is only temporary. Hubbert's peak for natural gas is estimated to occur only a couple of decades after the one for cheap oil.

There is also a possible fuel called methane hydrate, a solid that looks like ice but burns when ignited. Consisting of methane molecules trapped in a cage of water molecules, it forms when methane combines with water at temperatures close to the freezing point of water and under high pressure. Methane hydrate was discovered only a few decades ago. There are a number of theories about where it might be found, including under the Arctic permafrost, on the deep ocean floor and on the moons of Saturn, as well as how much of it there is and whether it can be mined and used successfully. Not much is known with certainty except that the stuff exists.

A huge amount of chemical potential energy is stored in the Earth in the form of coal, which is primarily elemental carbon. As is true of the other fossil fuels, to extract the stored energy from coal, each atom of carbon must combine with oxygen to become carbon dioxide, a greenhouse gas. But in addition to it inevitable carbon dioxide, coal often comes with unpleasant impurities such as sulfur, mercury and arsenic, none of which can be inexpensively extracted. Coal is a very dirty fuel (the dangerous concentrations of mercury that are found in the meat of swordfish and tuna originate in coal-fired power plants). Nevertheless, coal can be liquefied by combining it with hydrogen at high temperature and pressure, an expensive, energy intensive process that Germany used out of desperation in World War II. If we take our chances on fouling the atmosphere and turn to coal as our primary fuel, we are told that there is enough of it in the ground to last for hundreds of years. That estimate, however, is like the R/P ratio for oil. It does not take into account the rising world population, the determination of the developing world to attain a high standard of living and above all the Hubbert's peak effect, which is just as valid for coal as it is for oil. The simple fact is, if we turn to coal as a substitute for oil, the end of the age of fossil fuel, coal included, is sure to come to an end eventually.

Controlled nuclear fusion — energy obtained by fusing light nuclei into heavier ones — has long been seen as the ultimate energy source of the future. The technical problems that have prevented successful use of nuclear fusion up till now may someday be solved. Not in time to rescue us from the slide down the other side of Hubbert's peak perhaps, but maybe someday. Then the fuel — at least initially — would be deuterium, a form of hydrogen found naturally in sea water and lithium, a light element found in many common minerals (the nuclear reaction envisioned for fusion reactors is the fusion of deuterium and tritium, two isotopes of hydrogen. Tritium does not exist in nature, but the fusion reaction yields neutrons, which can be used to make tritium in a lithium blanket). There would be enough deuterium and lithium to last a very long time. However, the conquest and practical use of nuclear fusion has proved to be very difficult. It has been said of both nuclear fusion and shale oil that they are the energy sources of the future and always will be.

Nuclear fission, on the other hand, is a well-established technology. The fuel for this kind of reactor is the highly radioactive isotope uranium-235. The very word "nuclear" strikes fear into the hearts of many people — so much so that the utterly innocent imaging technique called nuclear magnetic resonance (NMR) by scientists had to be renamed magnetic resonance imaging (MRI) before the public could accept it for medical use. When the oil crisis occurs, the fear of nuclear energy is likely to recede, because of the compelling need for it. However, there will continue to be legitimate concerns about safety and nuclear waste disposal. Also, nuclear energy is suitable only for power plants or very large, heavy vehicles, such as ships and submarines. Do not look for nuclear cars or airplanes anytime soon.

Let us suppose for one euphoric moment that one more really big oil field is still out there waiting to be discovered. The largest oil field ever found is the Ghawar field in Saudi Arabia, whose 87 billion barrels were discovered in 1948. If someone were to stumble onto another 90-billion-barrel field tomorrow, Hubbert's peak would be delayed by a year or two, well within the uncertainty of the present estimates of when it will occur. It would hardly make any difference at all. That fact points up the sterility of the long-standing debate over drilling for oil in the Arctic National Wildlife Refuge (ANWR) in Alaska. If the ANWR is opened for drilling, its contribution to world supplies will be modest indeed. The best reason for not drilling there is not to protect the wildlife. It is to preserve the oil for future generations to use in petrochemicals, rather than burning it up in our cars.

Once Hubbert's final peak is reached and oil supplies start to decline, how fast will the gap grow between supply and demand? That is a crucial question and one that is almost impossible to answer with confidence. Here is a rough attempt at guessing the answer. The upward trend at which demand has been growing amounts to an increase of a few percent per year. On the other side of the peak, we can guess that the available supply will decline at about the same rate. The gap then would increase at about, say, 5% per year. Therefore, 10 years after the peak, we would need a substitute for something like half the oil we use today — that is a substitute for something like 10 billion to 15 billion barrels per year. Even in the absence of any major disruptions caused by the oil shortages after the peak, it is very difficult to see how an effective substitution can possibly be accomplished.

To be sure, the effects of the looming crisis could be greatly mitigated by taking steps to decrease the demand for oil. For example, with little sacrifice of convenience or comfort, we Americans could drive fuel-efficient hybrids rather than humongous gas-guzzlers. Some motion in this direction has in fact taken place, but it is trivial compared with the magnitude of the problem. There are countless other ways in which we could reduce our extravagant consumption of energy: Redesign cities, better insulated homes, improved public transportation and so on. Such changes are beginning to be made, but there are powerful interests — like the oil companies and the automobile industry and its unions — opposing them.

Before we turn to prospects for the future, a little summing up is in order: The followers of King Hubbert may or may not be correct in their quantitative predictions of when the peak will occur. Regardless of that they have taught us a very important lesson. The crisis will come not when we pump the last drop of oil, but rather when the oil that can be pumped out of the ground starts to diminish. That means the crisis will come when we have used roughly half the oil that nature made for us. Any way you look at it, the problem is much closer than we previously imagined. That is true even if we stretch things out a bit by harvesting difficult supplies of oil. Moreover, burning fossil fuels alters the atmosphere and could threaten the balmy, metastable state our planet is in. We clearly have some very big problems to solve.

3.1. Future Scenarios

So, what does the future hold? We can easily sketch out a worst case scenario and a best case scenario.

Worst Case: After Hubbert's final peak, all efforts to produce, distribute and consume alternative fuels fast enough to fill the gap between falling supplies and rising demand fail. Runaway inflation and worldwide depression leave many billions of people with no alternative but to burn coal in vast quantities for warmth, cooking and primitive industry. The change in the greenhouse effect that results eventually tips the Earth's climate into a new state hostile to life, such as exists on the planet Venus. End of story. In this instance, worst case really means worst case.

Best Case: The worldwide disruptions that follow Hubbert's peak serve as a global wake-up call. A methane based economy is successful in bridging the gap temporarily while nuclear power plants are built and the infrastructure for other alternative fuels is put in place. The world watches anxiously as each new Hubbert's peak estimate for uranium and methane makes front-page news.

No matter what else happens, we must soon learn to live without fossil fuels. Either we will be wise enough to do so before we have to, or we will be forced to do so when the stuff starts to run out. One way to do that would be to return to life as it was lived in the eighteenth century before we started to use much fossil fuel. That would require, among many other things, eliminating roughly 95% of the world's population. The other possibility is to devise a way of running a complex civilization approximating the one we have now, which does not use fossil fuel. Do the necessary scientific and technical principles exist?

One of the more difficult problems will be finding a fuel for transportation. One possibility is that advanced electric batteries will make battery-powered electric vehicles practical. In the past decade, batteries packing many times as much energy in a given volume as the batteries commonly used in cars have been developed for use in mobile phones and portable computers as well as in cars. There is no reason why even better advanced batteries cannot become the basis of a future means of transportation. Alternatively, the transportation fuel of the future might be hydrogen — not deuterium for thermonuclear fusion but ordinary hydrogen, to be burned as a fuel by old fashioned combustion or used in hydrogen fuel cells, which produce electricity directly. Burning hydrogen or using it in fuel cells puts into the atmosphere nothing but water vapor. Water vapor is a greenhouse gas, to be sure, but unlike carbon dioxide it cycles rapidly out of the atmosphere as rain or snow.

Hydrogen is dangerous and difficult to store, but so are gasoline and methane. Nature has not set aside a supply for us, but we can make it ourselves. There would inevitably be other problems associated with the large-scale use of hydrogen. For example, inevitable leaks to the atmosphere would be a threat to the Earth's protective ozone layer. However, let us suppose that those problems can be solved.

Still, you cannot get something for nothing. Hydrogen is a high-potential-energy substance; that is precisely why it is valuable as a fuel. So is the working fluid of an electric battery. But, thermodynamically speaking, hydrogen and batteries are not literally sources of energy. They are only means of storing and transporting it. That energy has to come from somewhere. Where will we get the energy, for example, to make hydrogen? Interestingly, one possible source is the potential energy stored in coal. There are existing processes that combine coal and steam to make hydrogen — and inevitably, carbon dioxide. Hydrogen gas is produced from a slurry of water and coal using a calcium oxide to calcium carbonate intermediate reaction. The calcium carbonate is then converted back into calcium oxide (to be used over again) and carbon dioxide. In principle, the carbon dioxide could be separated and stored — "sequestered" is the current buzzword. Where could it be sequestered? That problem has not been solved yet. And in any case the coal will eventually run out, whereas we are trying to think long-term here.

The interior of the Earth is heated by the decay of natural radioactive elements. In a sense, we live right on top of a vast nuclear source. Can we not use all that energy? We do, to some extent. It is called geothermal energy, and it is conveniently used for space heating in some places, such as Iceland. Using it to generate power is more difficult. The temperature of the Earth's interior rises with increasing depth, typically reaching the boiling point of water at a depth of about three miles. There are only a handful of places on Earth where a geothermal source rises to within drilling distance of the surface and can thus be used for power generation. Even in those places, using the heat to generate power often cools the source faster than its heat can be replenished. Geothermal energy will always be useful but probably never a major contributor. We should remember, though that without geothermal energy to heat the interior of the Earth, we would never have had fossil fuels.

There is a cheap, plentiful supply of energy available for the taking and like geothermal energy, it will not run out for billions of years. It is called sunlight. We now make very poor use of the sunlight that arrives at Earth. Farmers use it to grow food and fibers or textiles. A little bit is collected indirectly, in the form of hydroelectric and wind power. Here and there, solar cells provide energy for one use or another. But by and large, the Earth absorbs the solar energy that is not reflected back into space. We could learn to make better use of it along the way.

Sunlight is not very intense as energy sources go. The flux of energy from the Sun amounts to 343 watts per square meter at the top of the atmosphere, averaged over the entire surface of the Earth and over an entire year. By comparison, the continuous per capita consumption of electric power by Americans is 1,000 watts. Nevertheless, the solar power falling on the United States alone amounts to about 10,000 times as much power as even Americans consume. Both sunlight and nuclear power can be used to make hydrogen or charge batteries in a number of ways. There are chemicals and organisms that evolve hydrogen when sunlight is added. Sunlight makes electricity directly in solar cells. Electricity can also be generated by using sunlight or nuclear energy as a source of heat to run a heat engine — such as a turbine — that can generate electricity. By means of electrolysis, electricity can make hydrogen from water. There is not much reason to doubt that hydrogen or advanced batteries can serve our transportation needs. At present, nuclear technology is far advanced compared to solar for all these purposes, but that could change.

So, technically, scientifically, the means may exist to build a civilization that has everything we think we need, without fossil fuels. Thus, there may be a future for us. The remaining question is, can we get there? And if it is possible to live without fossil fuels, why wait until the fuels are all burned up? Why not get to work right now, before we do possible irreparable damage to the climate of our planet? The answer is we had better do that, or we will pay the consequences.

CHAPTER 4

THE ECONOMICS OF GLOBAL
ENERGY SECURITY

By now, it should be clear that nations across the planet face immense challenges in securing the energy needed for their economies. We shall now look at these challenges and discuss what lies ahead for these vulnerable nations, including their need for closer cooperation.

Some of the key issues involved in understanding national and international energy security today include the challenges stemming from energy–supply disruptions (Section 4.1.); long-term dangers in particular (Section 4.2.); issues related to growing dependency on natural gas (Section 4.3.); evolving relations with the Middle East (Section 4.4.); the prospect of a major global energy crunch (Section 4.5.); the role of markets (Section 4.6.); and finally, how new initiatives could address these issues (Section 4.7.).

4.1. Potential Dangers to the Economies of Nations of the World Stemming from Energy–Supply Disruptions

The dramatic geopolitical shifts following the 1991 collapse of the Soviet Union, the end of the Cold War, and the current global war on terror all constitute a major restructuring of the bipolar world order that had dominated the latter half of the twentieth century. No nation is immune to the repercussions of this radically altered geopolitical landscape. A convergence of new factors, ranging from the threats posed by such groups such as Al Qaeda to the sweeping engagement of the United States military in the Muslim world, including the invasions of Iraq and Afghanistan, has endowed the Middle East and Asian regions, including particularly East Asia and South Asia with a significantly enhanced strategic, political and economic importance. However, all nations, to a greater or lesser extent, are affected by these changes.

In fact, possible threats to energy security pose significant dangers to the economies of the United States, Europe, Asia-Pacific and other advanced and developing nations of the world, given their absolute dependence on energy supplies. The uninterrupted continuity of relatively inexpensive energy is crucial to the continued economic development of these nations, many of which rely upon imported oil, gas and coal. Energy security impacts them all, with potential downstream effects caused by energy supply disruptions.

Historical examples of these energy supply disruptions include the 1973 Organization of the Petroleum Exporting Countries (OPEC) oil embargo and the more recent examples of Russia cutting off supplies of gas to Ukraine and Belarus, with subsequent impacts in Western Europe, whose gas supplies fell significantly.

In terms of pure economics, the outlook for energy security in the world looks particularly troubling. Oil and coal consumption are on the rise world-wide, with serious implications for climate change; meanwhile, the demand for other energy imports is also steadily increasing. The imbalance of supply and demand affects both energy security and overall economic security, particularly in the developed economies; but this misalignment is also beginning to impinge on developing economies, particularly those of China, India and Brazil. Some energy experts have contended that the world's dependence on Middle Eastern oil may soon exceed 90%. Although newly discovered oil fields in Eastern Siberia and Central Asia, as well as new offshore oil found elsewhere and the widespread application of fracking techniques in the world promise to offer some short-term energy supply relief, the lack of an existing infrastructure to facilitate the transport of this oil poses additional political and economic challenges.

The dangers stemming from oil scarcity or shortage, combined with an ever-growing demand for energy, has the potential to strain relations among important energy-importing nations, including the United States, Europe, China, India and Japan. Conceivably these tensions could spark a new round of destabilizing regional or international conflicts, as these countries scramble for new supplies. They may of course opt to drill for oil off shore, dig for more coal, import liquefied natural gas (LNG), or build nuclear power plants, but none of these options is without potentially grave environmental consequences.

An even more immediate problem is the effect of oil-market volatility on the global economy, with rising oil prices that we have experienced at many times, including in the Spring of 2011. They would put intense pressure on the currencies of some oil-importing nations, along with the dangers of soaring current account deficits and weaker economic growth. These dangers threaten to curtail the record of remarkable growth that has served as the driving force for stability and development in many parts of Asia in recent decades. Indeed, the remarkable growth of Japan, South Korea, Taiwan and Singapore, and more recently China and India, can be explained, in part, on the basis of these nations' access to low-cost energy. While Asia may be the most vulnerable region, surges in oil prices would threaten other oil importers worldwide, including the United States and Europe. The danger for these economies lies in the impacts of high-energy prices on their international financial positions and growth, as well as on their domestic purchasing power. Further increases in energy prices could represent a major threat to their economies. On the other hand there have been and are still periods of low-cost oil. That situation threatens the economies of oil producing nations and renders some supplies such as Canadian tar sands uneconomic. There are severe problems when the price of oil goes up or goes down.

4.2. External Energy Supply Disruptions: Dangers over the Long Term

There are also major potential shortfalls of supply as traditional sources of energy of all types run out. While the "Hubbert's Peak", denouement for easy access to oil supplies has historically been applied to the United States particularly the lower 48, it is now a looming problem for the world as a whole. As we argued earlier in this book, the available petrochemical data suggest that we will run out of fossil fuels eventually. Is it that "we will run out of fossil fuels" or that "we will run out of new sources of fossil fuels" i.e., there will still be fuel to consume but no new sources on the horizon? The advent of Hubbert's Peak, the point at which we will have consumed half of all oil known to exist, and will likely exhaust the rest even faster, due to the ever-increasing demand for oil coupled with the decreasing discoveries of new fields. Even conservative calculations predict that the price of oil will increase significantly within the foreseeable future, a phenomenon we have already witnessed in 2011. We conclude by noting that, "No matter what else happens, we must learn to live without fossil fuels".

As noted above, we can expect to see major increases in China and India's demand for energy in the near term as well as new demands from rapidly modernizing nations including Brazil, South Africa and others. As of 2010, energy consumption in developing Asia (including China and India, but excluding Japan, Australia and New Zealand) had surpassed consumption in all of North America. Equally important, as of 2009, as shown below China has now outstripped the United States as the world's major air polluter. Over the next two decades, China and India alone are expected to account for over 40% of the increase in world oil consumption, 75% of the increase in world coal consumption and 45% of the increase in global carbon emissions.

According to the BP Statistical Review, North America's total consumption of oil increased from 18,474 barrels daily in 1984 to 22,826 barrels in 2009. Over the same period, the Asia-Pacific region saw the greatest growth in consumption, from 10,472 barrels in 1984 to 25,998 barrels in 2009, exceeding the consumption in North America (Table 4.1.).

Table 4.1.　Total consumption of oil in thousand barrels daily.

	North America	Asia-Pacific
1984	18,474	10,472
1985	18,535	10,500
1986	19,085	10,981
1987	19,598	11,277
1988	20,303	12,186
1989	20,503	13,018
1990	20,206	13,862
1991	19,908	14,474
1992	20,279	15,387
1993	20,586	16,124
1994	21,232	17,133
1995	21,150	18,212
1996	21,823	18,916
1997	22,276	20,020
1998	22,674	19,567
1999	23,286	20,518
2000	23,548	21,126
2001	23,571	21,282
2002	23,665	21,891
2003	24,050	22,671
2004	24,898	23,957
2005	25,023	24,331
2006	24,904	24,721
2007	25,020	25,462
2008	23,795	25,662
2009	22,826	25,998

In 2010, China vaulted ahead of the United States as the number one country for energy demand and there seems little doubt that its economy will demand ever more energy from a variety of sources in the decades ahead. The emergence of China and the Asia-Pacific region as the world's dominant energy consumer poses challenges to this region's future economic security. The region's current energy mix is determined by balancing its indigenous energy resources with policies that encourage a diversification of supply, development of strategic reserves, ongoing research for new domestic sources and new commercial approaches to overcome resource constraints. Overall, energy consumption in Asia is expected to surge in the next decade, as its major developing economies enter energy-intensive stages of economic development. For example, growth rates for automobile ownership in China, India, South Korea and Thailand, have exceeded 10% a year in recent years and China has now surpassed the United States as the largest consumer of automobiles in the world.

There is also a potential for continued conflict in Asia over oil and gas, pipelines, and so forth, with a possible renewal of the "Great Game" in Central Asia with several nations vying for its natural resources as happened in the nineteenth and early twentieth centuries. There is also potential conflict in Iran, and territorial conflict in the South China Sea between China, Japan and other nations over disputed islands that are believed to have substantial energy resources. These nations should instead be working cooperatively to address the challenges of energy security through conservation and joint research on renewable energy sources.

The world's rising long-term energy demand, led by China and India, has made Asia a dominant consumer of global energy supplies. Technological advances and a more sophisticated energy market have led to a growing reliance on natural gas and nuclear power relative to the region's traditional reliance on coal and oil. Nevertheless, coal and oil remain the region's primary energy sources, with the demand for oil expected to increase considerably over the next decade. As economic growth continues to be a top priority of the Chinese government, energy security naturally ranks as a top political and economic concern. China has been and will continue to try to ensure its energy security for the future by developing energy links both with its neighbors in the region and in Africa, Australia, Latin America and elsewhere. Over the next few decades, the nature and direction of industrial activity in China is expected to play a critical role in fueling the nation's accelerating energy demands. For instance, China is now the world's largest producer of iron, steel and cement, whose production requires building heavy-duty infrastructures throughout the country. Even with the use of the most modern production technology, these are particularly energy-intensive industries.

Access to energy sources is a critical security issue for the world as a whole, given the structure of international energy needs and expected future consumption patterns. Governments and security professionals continue to ponder traditional energy-security concerns such as safe access to sea lanes, reliable transportation, territorial conflicts and attendant environmental issues such as air and water pollution. Newer sources of energy, such as natural gas, a more sophisticated and integrated energy market, and newly emerging strategic relationships have introduced new energy security considerations across the globe.

A significant problem is that there is an incomplete recognition on the part of analysts and policy makers of the strategic importance of energy security. The current focus on energy security remains somewhat parochial, with a rather outdated reliance on the more traditional perspective of concentrating, for example, on the risks posed by instability in the oil-producing nations of the Middle East. Nonetheless, the Middle East still legitimately commands attention for three reasons: First, it continues to be the world's major energy supplier, with increasingly, a dominant market in Asia. Consumption of oil in Asia rose from 24 million barrels per day in 2000 to 34 million barrels per day in 2015.

Second is the instability rooted in the very nature of the Middle East regimes, and third is the region's ongoing potential as a staging ground for a possible new wave of terrorism. Important as these are, conditions in the Middle East are better understood as one of a series of potential challenges to global energy security. Let us consider some of these in turn.

4.3. Natural Gas Raises New Security Challenges

Many resource experts claim that natural gas will serve as the world's primary energy source in the future. Some even refer to the coming era as the "Gas Century". Of all current major energy sources, the worldwide use of natural gas is expected to grow most rapidly. Natural gas creates less environmental damage than other fossil fuels and it is considered a highly efficient means for generating electricity. In fact, much of the future growth in gas demand will be for electricity generation because combined-cycle gas-fired generators require shorter construction periods and are more efficient than other fossil fuel generators or nuclear power plants. Advances in technology have also raised the possibility of natural gas becoming a major energy source in the future. Abundant reserves exist in Central and Southeast Asia, so the potential for new gas resources and ready markets is very high. In short, assuming the appropriate regional pipeline networks are in place, the world's energy mix is likely to shift in the future toward greater reliance on natural gas. There are also newer technologies for extracting natural gas, such as hydrofracking in shale, although it raises important environmental concerns.

At the same time, natural gas carries its own set of concerns. With one quarter of global gas reserves, Russia clearly has the capacity to be a major energy supplier to the world, but there remains the issue of whether it is a reliable supplier given its previous cutbacks in gas supplies to Europe. Nevertheless, if these challenges can be successfully overcome, a growing international reliance on natural gas will reduce global dependency on oil, somewhat alleviating the importance placed on sea lanes as the lifeline to key economies.

LNG (Liquefied Natural Gas, a cryogenic or low temperature substance) also has great potential as a major energy source, but it too poses security and environmental concerns. High start-up costs and political risks hamper exploitation. The pipelines and supporting infrastructure needed to deliver the gas to markets require both bilateral and multilateral cooperation among countries whose perceptions of one another are characterized by suspicion on the one hand and a desire to cooperate for mutual benefit on the other. The more countries a pipeline must pass through, the greater the likelihood for contention and conflict, and the less likelihood that the pipeline will be completed. China is currently confronting this issue with natural gas being piped there from Russia and added pipelines being built.

The types of disagreements that could arise under these circumstances are not hard to imagine: Given the enormous costs associated with building a pipeline, which of the many proposed projects will be built? Who will build them? Who will pay? How will the pipelines be routed? The troubling absence of a comprehensive international energy security program is perhaps no more evident than in the vulnerabilities of key components of regional and global energy networks, including critical deficiencies in the transport of LNG and the exposed weakness of pipelines.

4.4. The Evolving Role of the Middle East

The world's voracious oil consumption today, its almost certain growth over the next decades, and the concentrated nature of the primary supply source in the Middle East, all combine to pose important security risks. Rapid economic growth and industrialization throughout Asia, particularly in India and China, will only intensify the demand for and dependence on Middle East oil. At the same time, Middle Eastern oil producers and exporters are becoming increasingly dependent on Asia for their volume growth in sales and revenue. From a political and security perspective, Asia's rapidly rising oil consumption, the anticipated growth in this demand and the concentrated nature of its supply source in the Middle East all pose several important questions.

First, is it in the world's energy security interest for any region to rely primarily on Middle Eastern oil? The history of political instability there certainly raises the possibility of large-scale disruptions in its ability to export oil. Although many energy-importing states have diversified their energy portfolios in terms of both energy sources and their suppliers, trends in recent years imply a continuing or even growing reliance on the Middle East to supply many of the world's oil needs.

Second, the increasing dependence on the Middle East heightens the strategic importance of sea lanes from the Persian Gulf across the Arabian Sea and the Indian Ocean, through the Straits of Malacca and nearby waterways, and finally across the South and East China Seas to China, Japan and South Korea. Currently, the United States naval presence in and around these waters ensures safe and open access to these sea lanes, demonstrating not only the importance of the United States role, but also underscoring how vulnerable energy-poor economies like Japan and South Korea are to any blockage of these waterways.

Third, East Asia and other oil-importing regions will likely gain a greater political and economic interest in the Middle East (and *vice versa*) as a natural spillover effect of this greater energy interdependence. China's deepening energy ties with Iran and Japan's moves toward closer relations with Abu Dhabi and Saudi Arabia are examples of the newly emerging Asia–Middle East relationships.

Fourth, the upward trend in the Asia–Middle East energy nexus is occurring at a time when the United States is reducing its dependence on Middle East oil and relying increasingly on oil from Canada, Mexico, Latin America and Africa, as well as from its own enhanced drilling. From a geopolitical perspective, growing Asia–Middle East ties over energy implies new strategic interests for Asia in the Middle East, which could create frictions in United States–Asia relations, especially with China, Japan, India and South Korea. A future scenario in which Asia is less readily supportive of the United States and Europe in employing economic sanctions or embargoes to influence political behavior in the Middle East and other regions, such as various African nations and North Korea, is quite possible.

At the same time, Asia's increasing reliance on Middle East energy may encourage the region's nations to behave in ways that enhance stability. For example, China may reconsider its policy of selling arms to potentially militant nations in the Middle East.

4.5. Possible Impacts of a Major Energy Crunch

What forms might a future energy crisis take? One possibility, echoing the oil embargoes of the 1970s, is that a major energy producer, either within or outside of OPEC, might be politically motivated to shut off supplies to the United States or other key oil-consuming nations.

Additionally, political upheaval or outright revolution inside a traditional energy producer, such as Saudi Arabia, Iraq, Iran, Kuwait and the United Arab Emirates (UAE), could reduce or shut off supplies.

Another danger concerns access to sea lanes, particularly given the trend toward greater oil consumption in Asia. Freedom of navigation remains a top priority, with America's longstanding commitment to ensuring open sea lanes a critical factor in regional and global stability as it relates to energy.

Compounding these issues is the impact of regulatory processes, environmental concerns, the overall vulnerability of the energy infrastructure, shifting strategic relationships, and potential territorial conflicts in the South and East China Seas.

4.6. The Role of Markets

It has been argued that today's more sophisticated energy market, with its privatization, competition and open markets, could provide the required resources, capital and infrastructure to deal with current energy risks. If so, it could lead governments to trust the market to effectively broker their energy needs, instead of relying on traditional energy security strategies, such as stockpiling.

It is true that governments no longer see their energy security problems exclusively from a supply–shortage perspective as they did during the 1970s.

Rather, the effects of market competition and more investment in energy sectors have led some countries to liberalize these sectors, making them more open to foreign investment. Still, less energy self-sufficient countries such as Japan and South Korea remain cautious about placing their energy security in the hands of the market. While energy companies may believe in the efficacy of markets, governments, particularly those that feel they have little margin for error, tend to be more reluctant about relying on the market to provide resources that they perceive as vital to national security, instead relying on command-economy approaches.

Authoritarianism of the past is now mostly central government policy oriented due to concerns of energy security where post-colonial state-hood and a tendency for governments to view the world from a more "realpolitik" conception of security may explain the particular reluc-tance of some nations to trust institutions or systems that weaken the control of the state. Differing levels of confidence among governments toward markets might suggest a potential role for regional institutions to mediate between governments and the market, ensuring a reliable supply of energy at reasonable prices to meet the world's growing energy demands.

4.7. Conclusion: Toward the Formulation of a Constructive Collective Energy Security Policy

Coordinated and farsighted public policy initiatives could address issues of how best to safeguard energy security in a number of ways.

First, efficiency and conservation in the use of energy resources must be taken more seriously, particularly in the major energy-using nations, including the United States, China, India and the countries of the European Union (EU). Energy management that includes reduced use of heating and cooling, along with energy-saving appliances and technologies, are of critical importance.

Second, the highest priority must be given to technological advances and new energy sources, especially renewable energy generation such as wind, solar, biomass and geothermal. This must be accompanied by a diversification of the energy-consuming nations' energy portfolios and greater use of storage and hybrid technologies.

Third, higher taxes will need to be assessed on final energy demands, with the proceeds used in part to develop new renewable energy supplies, but mainly to provide a direct dividend to the population. This could take the form of a "National Energy Dividend", such as the one that is currently provided in Alaska, based on oil production.

Fourth, the United States must publicly commit itself to a new energy program, embracing all of the above, much as President Kennedy committed the nation to putting a man on the moon. A major research program like that of the Manhattan Project to build the atomic bomb might be developed to identify new technologies for energy production and conservation.

Fifth, at the international level, energy security issues should be constructively addressed through diplomacy and collaborative interstate arrangements rather than through unwieldy and ineffective global commitments. International institutions could play a key role in this. Facing the common challenge of relying on external energy suppliers, the majority of the world's nations have significant incentives to cooperate rather than compete. Regional and global energy security requires a multilateral approach. The groundwork for this already exists, stemming both from converging national interests in the face of transnational threats and from multinational organizations already in place, such as Asia-Pacific Economic Cooperation (APEC) and the Association of Southeast Asian Nations (ASEAN), with similar multinational entities elsewhere in the world.

Sixth, the absence of governing regional structures has only exacerbated the various regions' vulnerabilities in today's volatile international environment. Although there has been some attempt to address these vulnerabilities through existing organizations, the regional states still lack the political will, military capability and most important, the experience to safeguard their energy security via concrete multilateral approaches.

Today, the major substantive security architecture in many regions is still a set of bilateral security treaties centered on the United States. It is time for a new regional and global approach to energy security. Such an effort could link economic cooperation to a regional security process and also build on the economic and political powers of each region. Energy security concerns may offer an important spur toward forging new regional arrangements, given the genuine level of cooperation and shared interests needed to peacefully secure energy supplies.

Despite the potential for regional cooperation on energy issues, the sheer scale and scope of diverging national interests has significantly impeded even early efforts at coordination. The absence of a recognized common goal is profound, making the pledges for joint strategic reserves and region-wide gas pipelines largely unfulfilled promises.

Seventh, it is clear that international institutions must play a key role in securing energy-resource security and stability. Financial institutions are particularly significant in this regard. The World Bank and other international development banks (IDBs) are charged with establishing and enforcing norms that favor cooperation. They could help provide financial security to countries that participate in pipeline projects by creating disincentives for states to take unilateral action against such a project.

Until regional institutions are created to address energy security concerns specifically, these international lending organizations may meet some of the demands posed by new challenges in the energy security arena. Such efforts may not necessarily foster regional cooperation on energy matters, but there is clearly a need for some type of organization to address a set of issues that are widely shared across the world.

Finally, establishing a robust and proactive international energy-security policy is essential to future global stability and sustainable development. But it remains to be seen whether the various world regions will be able to forge a collective and cooperative approach in the wake of other daunting challenges and demands, including an increasingly unstable unipolar world which is uneasily centered on the United States. In the broadest sense, energy security concerns have three facets. The first involves limiting the vulnerability to oil supply disruptions, which is essentially a short-term issue. The second pertains to the long-term concerns over the smooth functioning of the international energy system so as to ensure supplies to meet the rising demand at reasonable prices. Finally, a new dimension is that the production and use of energy must evolve in ways that promote sustainable development while minimizing damage to the environment. But even these efforts will amount to very little if the Hubbert's peak phenomenon kicks in and jeopardizes what little remains of the world's fossil fuel based energy supplies.

CHAPTER 5

ENERGY MYTHS AND A BRIEF HISTORY OF ENERGY

Here are a few common myths about energy and related matters:

- There is enough fossil fuel in the ground to last for hundreds of years.
- Four dollars a gallon is too much to pay for gasoline.
- Oil is produced by oil companies.
- When we do run out of oil, the marketplace will ensure that it is replaced by something else.
- Nuclear energy is bad.
- We can help by conserving energy. Otherwise, there will be an energy crisis.

None of these myths are correct. Some are outright false, and others express poorly something important that is correct. To get it right, we have to understand how it all works. Here we present how it all works.

Nuclear reactions inside the Sun heat its surface white hot. From that hot surface, energy in the form of light, both visible and (to our eyes) invisible, radiates uniformly in all directions. Ninety-three million miles away, the tiny globe called Earth intercepts a tiny fraction of that solar radiation. About 30% of the radiation that falls on the Earth is reflected directly back out into space. That is what one sees in a picture of the Earth, taken, say, from the moon. The rest of the radiant energy is absorbed by the Earth.

A body that has radiant energy falling on it warms up until it is sending energy away at the same rate as it receives. Only then it is in a kind of equilibrium, neither warming nor cooling. In any given epoch, Earth, like the moon or any other heavenly body, is in steady state balance with the Sun, neither gaining nor losing energy. That is the primary fact governing the temperature at the surface of our planet.

The rate at which the Earth radiates energy into space depends on its temperature. Because it receives only a tiny fraction of the Sun's radiation, it radiates much less energy than the Sun does, so it can balance its energy books at a temperature much cooler than that of the Sun. In fact, it can radiate as much energy as it receives with an average surface temperature of 0°F. Earth's radiation is not visible to our eyes and is called infrared ("below red") radiation, because its color is below the red end of what we are capable of seeing.

Fortunately for us, that is not all there is to it. If the average surface temperature of the Earth were 0°F, we probably would not be here. The Earth has a gaseous atmosphere, largely transparent to sunlight, but nearly opaque to the planet's infrared radiation. The blanket of atmosphere traps and reradiates part of the heat that the Earth is trying to radiate away. The books remain balanced, with the atmosphere radiating into space the same amount of energy the Earth receives, but also radiating heat back to the Earth's surface, warming it to a comfortable average temperature of 57°F. That is what is known as the greenhouse effect. Without the greenhouse effect and the global warming that results, we probably would not be alive.

There is a tiny but vital exception to the perfect energy balance of the Sun–Earth system. Of the light that falls on the Earth, an almost imperceptible fraction gets used up nourishing life. Through photosynthesis, plants make use of the Sun's light to grow. Animals eat some of the plants. Eventually, animals and plants die. Natural geological processes bury some of that organic matter deep in the Earth. As we have seen, that is how all fossil fuels are produced. So a tiny fraction of the distilled essence of sunlight is stored in the form of fossil fuels. Though the process of accumulating these fuels is agonizingly slow and inefficient, it has been going on for hundreds of millions of years and the Earth has built up a substantial supply.

As we saw earlier, many experts think there is enough oil and coal in the ground to last for centuries at the present rate of consumption. Among other fallacies, that view rests on the unstated assumption that the oil crisis will occur when the last drop of oil is pumped and likewise for coal and the other fossil fuels. The more sophisticated Hubbert's peak analysis tells us that we will get into trouble when we reach the halfway point. That is when the rate at which we can extract oil or the other fossil fuels starts to decline. But that is not the only fallacy in that rosy picture.

The present rate of consumption is the biggest myth of all. For one thing, we Americans consume fuel at five times the per capita rate as the rest of the world and the rest of the world wants in. For another thing, there is a powerful inverse correlation between per capita energy consumption and female fertility. The richer the nation, the higher the rate of fuel consumption and the fewer the number of children born. If the whole world is brought up to first-world status as quickly as possible, then some time later in the century there may be 10 billion people on Earth, living in relative comfort and burning lots of fuel. If, instead the third world remains in poverty, there may be a hundred billion people on Earth, living in poverty *and consuming the same amount of energy.* Either way, all the fossil fuel will run out a lot faster than predicted by the present rate of consumption. The fact that China and India are coming on-line as major car driving countries only makes matters worse.

Americans used to grumble about paying four dollars for a gallon of gasoline, but even at that price, gasoline was just about the cheapest liquid you could buy in the United States. Four dollars a gallon amounts to a dollar per liter. We pay twice that much for bottled drinking water. One consequence of cheap gasoline is that with 5% of the world's population, we consume 25% of the world's oil. Cheap gasoline is not part of the solution; it is a big part of the problem.

All of the oil pumped worldwide amounts to about 30 billion barrels annually. The oil companies refer to that as "production" but no oil company really produces a drop of oil. Instead they find it and extract it. That is the reason it is so cheap. Of course poking a hole in the ground and figuring out where to poke it does cost something. But can it really be true that this precious fluid that has taken the Earth hundreds of millions of years to accumulate is worth nothing more than the cost of pumping it out of the ground? Or have conventional economics, property rights and the rest somehow broken down here?

Speaking of conventional economics, economists firmly believe that when the oil starts to run out, the rising price will bring other, more expensive fuels to the marketplace. As we have already seen, the truth is a little bit more complicated than that. History shows that we do not react in an orderly, predictable way even to a temporary shortage of our precious gasoline. Whether we panic or not, the rate-of-conversion problem is likely to defeat us. Also, no other fossil fuel can replace the cheap oil that is the cornerstone of our civilization. Finally, if we do manage to burn up the heavy oil and other fossil fuels, the consequences for our climate cannot be predicted. All in all, we clearly have a serious energy problem.

Many people fear nuclear energy, but in reality, nearly all energy is nuclear in origin. The only energy sources we have are the Sun, which is a nuclear fusion reactor; natural radioactive elements in the Earth, which keep the interior of the planet hot; and man-made nuclear reactions in nuclear fission reactors and bombs. (The only exception to this rule is the energy of the tides, which arises from the rotation of the Earth. The only plant in the world to exploit this resource is found at the mouth of the La Rance river estuary in France.) Every other source of energy is derived from these sources. Of course, what people really fear is not the Sun or natural radioactivity buried in the Earth, but the nuclear reactors made by human beings. There have been some terrible accidents in nuclear reactors, but they are far less terrible than some of the accidents that have taken place in the history of coal mining or oil drilling. More than 100,000 men and boys died in the coal mines of England alone in the second half of the nineteenth century. By contrast, Chernobyl, the only nuclear reactor accident that caused a substantial number of deaths, is estimated to have caused 2,500 in all. While fuel for nuclear fission is also a finite resource, well run nuclear reactors are easily the safest and cleanest source of energy that is practical at this moment, provided we can find reliable ways to dispose of the nuclear wastes that result.

There is one other possibility: Energy from controlled nuclear fusion. Later, we will explore the technical differences between proposed nuclear fusion and the nuclear fission reactors of today. Nuclear fusion would use a fuel supply that is nearly inexhaustible and we know of no scientific principle that forbids it from working. As noted, however, it has proved remarkably elusive; in spite of billions of dollars invested, nuclear fusion has been 25 years away for the past 50 years. It seems unwise to bet the future of our civilization on nuclear fusion.

Surely, one way to help guarantee our future is by conserving energy. Surprisingly, however, it is not energy that we have to conserve. One of the most fundamental laws of physics says that energy is always conserved. Energy can change from one form to another, or it can flow from one body to another, but it can never be created or destroyed. We do not have to conserve energy because nature does it for us. For the same reason, there can never be an energy crisis. That does not mean we do not have a problem; it just means, we have not been describing the problem in the correct terms. There is something we are using up and that we need to learn to conserve. It is called fuel.

5.1. A Brief History of Energy

In the eighteenth century, heat was thought to be a fluid called caloric. Just as water flows downhill, caloric could flow down in temperature from a hotter body to a cooler one. And like water, caloric was neither created nor destroyed while it flowed. To use the jargon of modern physics, caloric was thought to be a conserved quantity. The caloric theory was rigorous and quantitative. A chunk of copper at a certain high temperature contained a known amount of caloric. If you put it into a container with a known amount of cool water, you could calculate how much caloric would flow out of the copper into the water and thereby predict with precision at what temperature the two substances would come to equilibrium. Nevertheless, a former American colonist named Count von Rumford found the caloric theory wanting.

Benjamin Thompson was born in Woburn, Massachusetts in 1753. Having spied and later commanded a regiment for England in the revolutionary war, he prudently went into exile, first in England and later in Bavaria. In the course of his career, Thompson promoted the use of James Watt's steam engine, introduced the potato into the common diet, invented a drip coffeemaker, and in 1791 was made Count von Rumford by the elector of Bavaria (Rumford was the name of what is now Concord, New Hampshire, his wife's hometown). He is best remembered today for pointing out in a scientific paper that boring out cannon barrels seemed to create quite a lot of caloric out of nothing. According to the caloric theory, that should not have been possible.

Count von Rumford's cannon barrels and many other observations would eventually blow the caloric theory out of the water. Caloric or heat would not turn out by itself to be a conserved quantity. It turns out to be just one of the possible forms of what we now call energy. Rumford and others during the first half of the nineteenth century tried to measure how much friction or other mechanical action would produce a given amount of heat. What they were looking for is what we now call the law of conservation of energy. It was discovered at least nine different times. When such a thing happens, credit for the discovery goes not to the person who discovered it first, but to the one who discovered it last — the one who discovered it so well that it never had to be discovered again. This person's name was James Prescott Joule.

Joule, son of a wealthy brewer, was born in Manchester, England, in 1818. Largely educated at home, he went off to Cambridge at the age of

16 to study under the famous chemist, John Dalton. After completing his education, he returned to Manchester where he built a laboratory in his father's house. Throughout his life, he supported his research out of his own pocket. In his most famous experiment, he arranged for a horizontal brass paddle wheel in a water tank to be turned by means of weights and pulleys. Weights of four pounds each descended a distance of 36 feet each at a rate of about a foot per second. Then they were hoisted up again, one after another, to keep the paddle wheel spinning. This procedure was repeated 16 times, after which the rise in temperature of the water was measured with a sensitive thermometer. Joule repeated the whole experiment nine different times, and did control experiments to determine the heating or cooling of the water by the atmosphere without the churning paddle wheel.

From the results of those experiments, he concluded that the amount of heat needed to warm a pound of water by 1°F — an amount now known as a British thermal unit or Btu — was equivalent to the amount of mechanical work required to lift a weight of 890 pounds through a distance of one foot. He achieved similar results in three more experiments: A magneto-electric experiment, another that involved the cooling of air by expansion, and another that measured the heating of water by constricting its flow in narrow tubes. Averaging the results of all of these experiments, he arrived at a value of 817 pounds lifted through one foot as the equivalent of one Btu. The accepted value today is 775 pounds. Notice that it requires an enormous weight lifted through one foot to equal a Btu. Just stirring your coffee will not heat it much.

A word about units: one *calorie* (a name left over from the old caloric theory) is the amount of heat needed to raise the temperature of one gram of water by one degree Celsius (1.8°F). A food calorie is actually a kilocalorie, the energy equivalent of 1,000 heat calories. Mechanical work is measured in units called *joules* (by no coincidence at all). A joule is the amount of work done in lifting a weight of one newton (about a quarter of a pound) through a distance of one meter. Thus, one calorie has the same energy as 4.2 joules. A Btu is equal to just about 1,000 joules.

It was perfectly clear to Joule that once the water had been warmed it made no difference whether that had been done by flowing caloric from a warmer body, churning a paddle wheel, or any other means of causing mechanical friction. The warmer water had an increased quantity of something he called *vis-viva*. We call it "energy".

The law of conservation of energy is one of the most important of all the laws of nature. Heat and work are the means by which energy can be transferred from one body or system or part of the universe to another.

Energy can exist in a number of forms. Among the most important are kinetic energy and potential energy. Gravitational, chemical and nuclear energy are all forms of potential energy.

Kinetic energy is the energy of motion. A car rolling down the street or a bowling ball rolling down the alley has kinetic energy because it is moving. When you step on the brakes to stop your car, the car's kinetic energy is turned into heat in your brake pads. Being an ardent thermodynamicist, I own a gas/electric hybrid car. When I (DG) step on the brakes at least part of the kinetic energy of the car's motion goes into recharging the car's batteries.

Even if a body is at rest, the atoms and molecules that make it up are undergoing constant random motions and therefore have kinetic energy. The absolute temperature of a body is proportional to the average kinetic energy of its atoms and molecules regardless of whether the body is solid, liquid, or gaseous. The hotter it is, the faster its atoms and molecules are jiggling around. The energy of the random motions of atoms and molecules should properly be called thermal energy, but often we loosely refer to it as heat.

A word here about temperature units: Absolute temperature is measured in Kelvins, which are the same size as degrees Celsius, but instead of placing the zero at the freezing point of water, the Kelvin scale starts at absolute zero. Absolute zero is the lowest temperature possible — the temperature of a body from which every last bit of moveable energy has been drained. This occurs at $-273°C$, or about $-459°F$.

Just as kinetic energy is the energy of motion, potential energy is the energy of position. For example, when Joule hoisted his weights 36 feet above the ground, he was doing work on them and thereby endowing them with gravitational potential energy. They had that potential energy by virtue of their position, 36 feet above the ground. As the weights descended, their potential energy was converted to the kinetic energy of the paddle wheel and the water, and ultimately to heat or thermal energy in the water.

Let us follow the energy through a sequence of ordinary events. You do some work when you lift a weight off the ground. The energy to do that work came from the sugary cereal you had for breakfast. The number of food calories (or kilocalories) was written right on the box. The weight now has potential energy. You drop the weight. Its potential energy, under the influence of gravity, immediately starts turning into the kinetic energy of the falling weight. By the time it reaches the ground, the potential energy is gone, having been replaced by an equal amount of kinetic energy of the falling weight. It hits the ground with a bang, and an instant later, everything is at rest.

What happened to your work? You are not the sort of person who feels your work is unappreciated, but, after all, energy is supposed to be conserved. Where did it go? The bang, when it hit the floor, is a good clue. There was a shock wave, propagating through the air and through the floor, both of which bounced around and eventually settled down as a slight increase in the random motions of the atoms and molecules of air, floor and everything else in the room. So, the net result of the whole sequence of events is that the food energy in your breakfast cereal has turned into useless heat.

There are countless examples in nature and in common experience of mechanical processes that seem almost but not quite to conserve a combination of potential and kinetic energy. For example, a pendulum, as it swings through its arc, trades the kinetic energy of its motion at the bottom of its swing for the potential energy of increased height as it comes to momentary rest at the end of its swing. But if we set a pendulum in motion and let it swing back and forth by itself, the motion gradually seeps out of it and it comes completely to rest. We say that friction in the pivot and in the pendulum's motion through the air gradually turn its potential and kinetic energy into heat. In other words, exactly the same amount of energy, we initially gave the pendulum, winds up as the jiggling motions of atoms and molecules.

Of course, we can say that, but how can we be sure that it is true? How can we know that it is not just a physicist's fiction for covering up an embarrassing disappearance of energy? The answer lies in Joule's ingenious experiments. When the organized energy of the swinging pendulum turns into heat, the energy does not vanish without a trace. It can still be observed in the form of a slight increase in temperature. In the case of the pendulum, or of the weight you dropped on the floor earlier, the increase might be too small to measure accurately. But Joule devised an experiment in which enough mechanical work was dissipated to allow the temperature increase to be measured with reasonable accuracy. Using it he showed that a given amount of mechanical energy — that is, organized as kinetic and potential energies — always turned into the same amount of heat. Thus, there is something that is conserved, always and everywhere. That something is what we call "energy".

When I (DG) taught this subject to the freshmen at the California Institute of Technology, I always used an impressive demonstration. A bowling ball was suspended by a long cable from the high ceiling of the lecture hall. I stood at one end of the stage, with the bowling ball held snug to my nose and the cable stretched taut. Then I would release the bowling ball to do its long pendulum swing across the width of the stage and back. As the ball went through its long swing there was plenty of time to remark to the class that this was an affirmation of my belief in the law of conservation of energy. If this one time the law were violated and after swinging 20 feet across the room and 20 feet back, the bowling ball arrived just an inch higher than it started, it would spoil my whole day. I would remark too, just how unpleasant it was to see a bowling ball rush towards you with nothing to protect you except gravity and the law of conservation of energy. Of course, it always finished its swing not at my nose where it started, but a few inches short because of friction and air resistance. It is a good idea to be careful not to lean forward when doing this demonstration.

Energy can be stored in chemicals. Food energy is one form of chemical energy. Fuel is another. Atoms are bound into molecules, where, because of their positions relative to one another, they have a certain potential energy. If they can get together with other molecules having different kinds of atoms, they might be able to arrange themselves into different kinds of molecules that have less potential energy. The excess energy is liberated to be used for other purposes.

Because oil and natural gas are crucial to the subject of this book, let us take them as examples. Oil and natural gas are made up of hydrocarbons — that is molecules that have a varying number of hydrogen and carbon atoms bound together. But they are bound rather loosely. If they are mixed with oxygen — or with air which is 20% oxygen — combinations that are more tightly bound are possible. The hydrogen atoms can detach from their molecules and hook up with the oxygen to form water, which is a very stable molecule. Likewise, the carbons can combine with oxygen to form carbon dioxide, which is also very stable. When a molecule is very stable — in other words very tightly bound — it has little or no further potential energy to give up. Because water and carbon dioxide have much lower potential energy than the fuel molecules, much energy is given off in the form of heat. That is what happens when fuel burns. The fuel will not burn, however, if it is merely mixed with air. The fuel molecules must unbind before the atoms can rebind in more favorable combinations. Something has to start the process going. That is what the spark plug in your car is for and that is why it takes a match to start a fire. Once burning begins, the process can produce plenty of heat to keep it going.

The nuclei of atoms also possess potential energy. On the periodic table of the elements, all those elements lighter than iron have nuclei that can recombine into other nuclei, which are less than the sum of their parts — that is, the final nucleus has a mass when at rest that is larger than the mass at rest of any of its constituent nuclei, but less than the sum of the rest masses of its constituents. The missing mass typically becomes heat, in accord with Einstein's famous formula. On the other hand, the nuclei of atoms heavier than iron can break apart, also leaving products with lower total rest mass. According to Einstein's formula, the excess rest mass of a nucleus is a form of potential energy. Iron nuclei have the lowest potential energy of any element on the periodic table.

When light nuclei combine, the process is called *fusion*. Fusion reactions provide the energy of the Sun and also the hydrogen bomb. That is the kind of nuclear reaction, we have not yet learned to make use of in reactors on the Earth. When heavy nuclei disintegrate, that process is called *fission*. That was the type of reaction that took place in the earliest nuclear bomb and it is the kind in all man-made reactors that are currently practical. Thus nuclear energy can be released by either fission or fusion, but, unfortunately, we have not yet learned how to control nuclear fusion.

CHAPTER 6

THE ECONOMICS OF ENERGY
AND CLIMATE CHANGE

6.1. Introduction

Economic factors are closely intertwined with energy and climate change issues, and this chapter will address some of these important interactions. As noted earlier, energy is a primary basis for all modern economies, which are largely fueled by the carbon-based resources of oil, coal and natural gas — the fossil fuels that are currently our main sources of energy. Clearly, any changes in these resources will have profound impacts on advanced economies, such as those of North America, Western Europe, Australia and Japan, as well as on the newer and emerging major economies of China, India, Russia, Brazil, Eastern Europe and South Africa. Changes in the availability of fossil fuels will also have significant impact on other developing economies of Africa, Asia and Latin America. At the same time, the development of these economies will have important impacts on energy resources and their evolution and impact. In addition, as noted below, the economics and climate issues also are tied in with political and human rights issues as became particularly evident in 2011 with the "Arab Spring" uprisings and the situation in the Middle East.

Concerns over climate change stem largely from the widespread burning of carbon-based energy resources of coal, oil and natural gas produced from the decay of primeval plant and animal matter. The burning of oil and coal produces carbon dioxide (CO_2) and other greenhouse gases that, as discussed earlier, have profoundly negative environmental impacts. The burning of natural gas, which is mostly methane, leads to the production of water and CO_2 that also have negative environmental impacts but only about half of those stemming from the effects of burning coal per unit of electricity produced.

In this chapter, we discuss several of the economic approaches that have been proposed to deal with this challenge. We discuss the economics of energy in terms of both the role of energy in the economy and the reciprocal influence of the economy on the energy sector.

The world is now in the beginning stages of a very important transition from artificially low-cost carbon-based fossil fuels, upon which it has been relying for many decades, to higher-cost alternatives, specifically to alternative fuels that do not lead to emissions that can cause climate change. This transition is likely to involve substantial challenges, including energy shortages, increase in the cost of various fuels, and a worldwide search for new energy sources. Also, the outcome of this transition will have profound impacts on every fuel-dependent national economy as well as on individuals, industry, governments and a wide range of organizations. The days of cheap energy that have fueled economic growth worldwide for many decades now are probably over, and the transition to new energy resources will be a costly and demanding process.

One might say that the price of fuel has been kept artificially low by the United States and its Allies and at great costs: Monetarily, politically and in terms of human rights. That is, for example, the United States and its allies have paid many billions — and maybe trillions — of dollars to keep in power the various regimes in the Organization of Petroleum Exporting Countries (OPEC) countries and Middle East by bolstering "friendly" regimes with loans, military sales and support, trade and other political favors. These costs are in addition to the trillions of dollars spent in wars in these regions. The human lives and human rights costs also have been high due to the United States support of totalitarian regimes, dictators, other suppressive governments, as well as the losses in wars. The Arab Spring and continuing uprisings in 2011 emphasized the widespread nature of these practices and their political, monetary and human costs.

Many people now recognize that climate change is a real and pressing issue facing all nations. The awarding of the 2007 Nobel Peace Prize to the UN's Intergovernmental Panel on Climate Change (IPCC) and former Vice President Al Gore for their work in this area, as well as the awarding of an Academy Award Oscar to Gore for his 2006 documentary film "An Inconvenient Truth", has certainly contributed to the widespread recognition of this issue. An important fact is that nearly all scientific authorities worldwide now regard environmental degradation due to climate change as a serious, even urgent, problem facing the planet. Scientific and technical solutions to this problem are being actively sought on numerous fronts. Closely intertwined with these energy and climate change issues are related economic (and political) issues.

The far lesser but gradually growing reliance on alternative and renewable fuels, including hydro (water), wind, solar, geothermal, biomass and nuclear power, has a key role to play and the use of these fuels could and should grow substantially in the future. Economic policies could play a major role in promoting greater use of these fuels thus reducing reliance on fuels that generate greenhouse gases leading to climate change.

6.2. The Stern Review

The authoritative Stern Review on "The Economics of Climate Change", prepared in the United Kingdom under the direction of the distinguished British economist Lord Nicholas Stern, was a landmark in alerting the world to the adverse consequences of climate change and to the role of human activity in causing this change. The review was released in 2006 and it has played a major role in all subsequent work in this area. Its overall conclusion was that, "The scientific evidence is now overwhelming: Climate change presents very serious global risks and it demands an urgent global response".

The review discussed the scientific evidence for human-caused climate change and its impacts. It also discussed the policy challenges that will inevitably be involved in transitioning to an economy based primarily on renewable, low-carbon fuels. It noted the urgent need for global action and international cooperation in "creating price signals and markets for carbon, spurring technology research, development and deployment and promoting adaptation, particularly for developing countries".

The review, considered the risks of persisting with the "business-as-usual" (BAU) approach and concluded that it would have irreversible adverse impacts on all economies of the world. It noted that the dynamic feedbacks in climate systems would amplify the effects of climate change and emphasized that these feedbacks have a high probability of producing a significant rise in average global temperature, resulting in damaging and irreversible consequences for the planet. It further noted that these consequences include threats to water supplies, food production, health and the overall environment. Greenhouse gases would lead to melting glaciers, rising sea levels, declining crop yields and impacts on global ecosystems that would, in fact, accelerate as the world becomes warmer.

The review, further noted that these impacts would particularly affect developing countries and the poorest people on the planet. Thus, it concluded that there is a strong case for mitigating these risks through new energy technologies and new economic policies. Such policies include reducing the demand for emissions-intensive goods and services as well as switching/shifting to low-carbon technologies for power, heat and travel via a carbon tax that would put an appropriate price on carbon, as discussed below. Other policy suggestions include direct regulation, trading schemes such as cap and trade, behavioral changes and the use of new technologies, also discussed below. The review's authors estimated that the annual costs for stabilizing the atmosphere by 2050 would amount to 1% of global GDP, concluding that this is a level that is "significant but manageable".

The Stern Review emphasized that there are three elements of policy that are essential for the mitigation of global warming: A carbon price, discussed below as a carbon tax; a technology policy, discussed in a later chapter on research on innovative technologies; and the removal of barriers to behavioral change. All of these elements would be needed for a successful program of mitigation. It further noted that there is still time to avoid the worst impacts of climate change if strong collective action starts now and that such action should build on the principles of effectiveness, efficiency and equity.

6.3. Energy Economics

Here, we treat the economics of energy in terms of such standard economic concepts as demand, supply, production, consumption and cost. In today's world, the demand for energy is mainly from industry, both public and private, although there is also a smaller and not inconsequential demand from households and governments. Since energy is fundamental to the operation of industry, its demand is relatively price inelastic, which means that changes in the prices of various energy sources have relatively little effect on their demand. In other words, prices can vary over a wide range, but demand stays relatively constant. The demand for energy is also income inelastic in that the users of energy, mainly industry, do not change their use of energy over wide ranges of their income.

Economists refer to the demand for energy as a "derived demand", meaning that it stems from the demand for the products produced by this energy. Thus, the demand for coal is largely a derived demand stemming from the demand for electricity since most electricity is produced in coal-fired power plants. In general, the demand for energy has been steadily increasing, and this is expected to continue given the rising demand for products that use it, including heating and transportation — the primary uses of oil and gas — and electricity — the primary use of coal. Estimates of future world demand for these fuels depend on projections of economic growth, which are intrinsically subject to uncertainty. Nevertheless, it is clear that a major challenge will be to find new energy sources, as well as devising new technologies to enhance their production and use to meet rising global energy demand.

The supply of energy depends on its production, which varies depending on whether it is in the form of fossil or renewable fuels. The supply of fossil fuels is subject to what economists call "increasing returns to scale", meaning that doubling the various inputs — including capital, labor and energy — in a commodity's production more than doubles its output. As a result, most of these fuels' production is concentrated in the hands of a few, very large producers that have acquired substantial political as well as economic influence. As these producers expand their output to meet rising demand, they have been compelled to invest increasing resources in the quest for and production of fossil fuels. Oil, for example, which was once relatively easy to discover and extract, is now considerably less plentiful and, relatively, much harder and costlier to find and extract. Oil rigs have had to go offshore and drill to greater depths in more and more remote and risky locations, with all the attendant potential environmental hazards. We saw this in the 2010 Gulf of Mexico Deepwater Horizon massive oil spill. Other such spills, even much larger ones, could well occur worldwide, such as in the Arctic region as drilling expands there. A spill in this region would be a much more serious problem and one much more difficult to treat than the one in the Gulf of Mexico.

Similar considerations apply to coal. Although, the innovation of mountain top removal has partially changed the economics of coal production, this development has come at the cost of negative and even disastrous environmental impacts. Natural gas is relatively plentiful worldwide, but, as with oil, producers have had to obtain it from increasingly remote locations. This has required them to make major investments in pipeline construction and technology, thus raising its cost. These costs are passed on to consumers, as is the case with the costs of other technologies for energy production.

The recent development and expanded use of the process of hydrofracking to release natural gas from shale has opened up new sources for this important energy resource and is one of the most important recent developments in the economics of energy. However, the extraction process itself is very expensive and it creates its own major environmental problems and risks. In recent years, hydraulic fracturing, commonly known as "fracking", has become as popular as it is controversial. The technique of fracking entails the extreme high-pressure injection of fracking fluid underground to create new cracks and crevices that are thought to ultimately yield higher quantities of gas and petroleum. The controversy over this process stems from the concern that by forcing such immense pressure underground, the process can inadvertently induce environmental impacts, including earthquakes. This concern has been corroborated where fracking has been used in areas not known for seismic activity and subsequent earthquakes have occurred.

As regard to renewable fuels, there are very substantial fixed costs in developing these fuels for operational use, whether in the form of dams for hydro (water) power, nuclear power plants, solar collecting arrays, turbines for wind farms, etc. These costs have to be repaid in the process of supplying the associated energy; and the payback period, the time when these fixed costs are finally covered, tends to be very long indeed.

As noted earlier, energy consumption is relatively straightforward, dominated by industry and typically price and income inelastic. Non-industrial consumers typically ignore the energy sector except when their attention is riveted by an energy crisis on the scale of the 1970s oil embargoes or the shut off of gas from Russia to Ukraine and to other parts of Europe in January 2006 and again in January 2009. Massive electricity blackouts, such as that in Hydro-Quebec in 1989 due to a solar storm disrupting the Earth's geomagnetic field, while rare, also make people at least fleetingly aware of the importance of energy in their lives. More severe low-probability but high-consequence events in the form of solar activity, hurricanes and earthquakes with tsunamis can destroy infrastructure and result in longer-term serious aftereffects including the loss of energy, transportation, food, potable water and the widespread disruption of banking and finance, communications, GPS and navigation systems. Many of these aftereffects were seen in the earthquake and tsunami disaster at the Fukushima Daiichi nuclear power plant in Japan in March 2011.

Most oil is produced in the developing world in such nations as Saudi Arabia, Iran, Mexico, United Arab Emirates, Iraq, Kuwait, Venezuela, Nigeria, Brazil, Angola, Algeria, Libya, Kazakhstan, Azerbaijan and Indonesia, many of which belong to the OPEC cartel of oil exporting nations. This cartel plays a key role in setting the price of oil, particularly through the actions of Saudi Arabia, which is so large a player in this market that it can affect the oil price by simply changing its supply. At the same time, some advanced economies are also major oil producers, including Russia, the United States, Canada, Norway and the United Kingdom, none of which belong to OPEC. Most of this oil is consumed in refined form in the advanced econo mies of the European Union (EU) and the United States, with China and India rapidly increasing their consumption.

Russia is a major producer and supplier of natural gas, and it has proposed the formation of an OPEC-like cartel for this gas. It also has exerted leverage on nations to which it exports its gas by shutting off its supply or threatening to do so. Its largest corporation overall, by far, is Gazprom, the monopoly that produces, transports and supplies natural gas and which has now also become a major producer and distributor of oil.

Economists refer to the impact of energy use on climate change as an "externality" by which they mean the direct effect that one economic agent has on another without going through the price-market system, i.e., a direct impact of one agent's action on the overall situation of another. Burning fossil fuels is an example of such an externality since it has important impacts on global warming and climate change that do not involve any market transactions.

One of the main approaches that economists use to study policy issues, including those related to energy, is cost-benefit analysis. The benefits net of costs of a policy initiative are determined over a future period of several years. They are then put on a comparable basis over time by discounting the future benefits net of costs at an appropriate discount rate and aggregating them to determine the overall present value of alternative policies. This approach has been applied to the methods discussed below for mitigating climate change and it has been found that they all make a net positive contribution to social welfare.

6.4. How Economic Tools can Mitigate Climate Change

Economic tools can mitigate climate change impacts in substantial and significant ways. We emphasize, however, that these are of limited utility. What is most urgently needed is first, scientific and technical research that would yield new approaches for addressing the climate change problem (as discussed in the next chapter) and, second, new and insightful types of economic analyses. In this context, it might be useful to recall that, historically, after the steam engine was invented, some people suggested shutting down the British patent office in the belief that there was nothing else left to invent. In Chapter 7, we will discuss the need for major research projects aimed at developing innovative energy-producing technologies that will not have the adverse effect of causing climate change.

Today, only about 10% of United States electricity is produced from renewable sources, with the balance coming from coal (45%), natural gas (24%) and nuclear power (19%). Some possible alternative and newer sources of energy that might replace the heavy dependence on coal include:

- Hydro power, but building dams is very time consuming and expensive and most of the rivers that are easy to dam have already been dammed up.
- Wind power, using large wind farms. The United States was an early leader in this field, but it has now been overtaken by nations in Europe and Asia. Wind power is a useful technology, but there is a limit to how much energy can be generated this way and wind farms, as their critics are swift to point out, also have their own environmental drawbacks as well as down time when the wind is not blowing.
- Greater use of natural gas, a source that many see as the energy of the future. Natural gas is abundant and much cleaner than the other fossil fuels. At the same time, however, it requires major investments in infrastructure, including pipelines and plants to produce it in liquified form (liquified natural gas, i.e., LNG), as well as ships to transport this LNG and plants to convert it back to a gaseous state. Hydrofracking, a technology that extracts natural gas from shale rock, has been growing in its use and has great potential for generating clean natural gas. Indeed, it represents one of the most important recent developments in the area of energy production, but, as already noted, it involves major environmental problems and risks.
- Greater reliance on nuclear power, which has been in use for many years and was once considered a major future source of electricity in the United States and other nations. It has the advantages of using a currently abundant fuel source, uranium and generating electricity without producing any greenhouse gases, but there are serious problems with this source of power. These include the enormous capital costs of building nuclear power plants and the challenges and risks that their operations entail. There is the problem of safely storing nuclear waste,

both in light of the hazards that exposed nuclear waste poses to life and the environment and with regard to the possibility that discarded radioactive material could be used to make dirty bombs, with the radioactive material dispersed with a conventional explosive or even to make nuclear weapons themselves. The world has also seen dramatic evidence of the damage that can result from major disasters at nuclear power plants, most recently at the Fukushima Daiichi plant in Japan following an earthquake and tsunami. Earlier, such incidents include the 1979 nuclear accident at Three-Mile Island in Pennsylvania, and the 1986 catastrophe at Chernobyl in the former USSR, which caused widespread radioactive contamination in Ukraine and Belarus. Some European nations, including Sweden and Germany, have decided to phase out their nuclear power plants, incurring substantial costs in decommissioning them. It should also be noted that uranium reserves are projected to reach their own Hubbert's peak within the next several decades and thus cannot be considered a limitlessly plentiful resource.

- Fuel from agricultural products, including ethanol from corn. Interest in this fuel source has been growing rapidly in the United States and has been a bonanza for corn growers, but many question the use of food to create energy rather than to feed people. Others question if the energy costs associated with producing ethanol are greater than the energy outputs. There are also environmental problems stemming from this way of producing energy.

- Biomass, including waste products from the forest industry and farms, which is currently under development and useful but is relatively small in scale.

- New sources of oil. While some have called for additional oil production, involving drilling deeper or farther offshore and in increasingly remote locations, including the Arctic, others have countered that these new sources soon would be depleted. For years, there have been calls to open the Arctic National Wildlife Refuge (ANWR) in Alaska to drilling and to use the oil in the Strategic Petroleum Reserve, an emergency fuel store of oil maintained by the United States Department of Energy. This drilling, however, may contribute to adverse environmental effects. One might argue that if this additional drilling were viewed as a temporary "bridge" measure until more environmentally friendly alternatives are developed, it might be acceptable. However, given the major financial investments in the initial drilling and development of a site, it may not be realistic to expect the developers to terminate the drilling.

6.5. Limits on Emissions, a Carbon Tax and a Cap-and-Trade System

Economists refer to a situation in which prices do not reflect the true scarcity of a product as one of "market failure", calling for imposing a tax or using regulation of some type. Fossil fuels are a major example. Economists have studied two market-based policy approaches to mitigate the problem of climate change: A carbon tax and a cap-and-trade system.

Both of these would work by imposing a price on emissions of greenhouse gases, economists have also considered other approaches that would directly limit such emissions.

In fact, the most straightforward way to control greenhouse gas emissions that cause global warming is simply to place direct limitations on them. These are already currently widely used in the EU and include restrictions on auto and truck exhaust and on smokestack emissions. This method of control does not involve market transactions, although it could be interpreted as imposing a substantial penalty price for exceeding the mandated limits. A further use of direct regulation would be a requirement that a certain share of a country's electricity generation, perhaps 25%, be obtained from renewable sources such as wind or solar power. However, economists, as discussed below, have also proposed indirect alternatives. They include a carbon tax and a cap-and-trade system.

Some conservative economists and political leaders throughout the world have proposed using none of these approaches. Instead, they argue for total reliance on the free market system, with no intervention on the part of the government. Of course, we have had a relatively free market in the energy area for many years and it has not solved the problem of greenhouse gases causing climate change and in fact has only permitted this situation to get worse. Thus, many have recognized that there is a need to intervene, with new public policy tools being required in this area. These tools will undoubtedly have an impact on how individuals and firms use energy and will likely be of long duration, so any implemented policy initiatives must be seen as both reasonable and necessary.

Renewable or clean electricity standards would involve a greater reliance on energy sources that produce little CO_2 or other greenhouse gases — sources such as hydro or wind power. Clean electricity standards would require that a certain share of electricity generation come either from such sources or from non-renewable sources that reduce emissions, such as nuclear power plants, coal-fired plants and those using natural gas that capture and store emissions, a process still under development. Such standards already exist in a majority of U.S. states, and they have been the subject of a study by the Congressional Budget Office (CBO) that explores what would be involved in establishing such standards for the nation as a whole. Currently, only about 10% of electricity in the United States is produced from renewable energy sources.

According to the CBO study, under a renewable energy standard, the federal government would give credits to producers of electricity who use renewable energy sources, including hydropower, wind and biomass, as opposed to fossil fuels. The producers could then sell these credits to the highest bidder, the bidders being fossil-fuel producers looking to buy offsets from other clean energy producers, with these offset credits meant to encourage producers to shift away from fossil fuels toward clean energy sources. Those utilities that generate electricity would comply with the policy either by using credits that they have received or those that they would buy from other generators that use qualified sources. The producers' demand for these credits would encourage firms generating electricity to comply with the standard, thereby encouraging them to produce more from renewable sources. If these credits are traded freely, then the market would determine the least expensive way of achieving the desired increase in renewable electricity generation. Under a clean-energy standard, electric utilities would have to submit these credits to the government, with each representing a megawatt hour of electricity generated from a qualified clean source. Utilities that are using fossil fuels would buy the credits that they need from those producers of electricity, who receive credits from the government for the qualifying electricity they produce. The demand for these credits would encourage the producers to make clean electricity that would generate credits for them. If the credits were traded in a free market, the market would determine the least costly way of achieving the desired increase in clean or renewable electricity generation.

The widely discussed carbon tax is a price on greenhouse gases that would impact both consumers and producers. It is a tax that is imposed on the price of energy products that produce greenhouse gases, including gasoline and home heating oil. Such a tax would raise the price of energy use and thus should reduce its use. Many arguments support levying such taxes, and many economists have endorsed them. The taxes would create positive incentives for consumers of energy to reduce their usage and they would also encourage the development of more efficient carbon-reducing technologies.

A cap-and-trade system is advocated by some economists and by some political and industry leaders and it is another approach to controlling greenhouse gases. It is supposed to involve setting a limit or cap on emissions for each source of pollution and then letting the producers of emission-creating products, including electricity generators, oil producers and importers, and natural gas processors, trade their rights to the limited emissions in a market specifically designed to carry out such transactions. It is supposed to be a market-driven system that would set a ceiling on global warming pollution while having the flexibility of allowing companies to trade permits to meet it. It would create a direct incentive to cut emissions, and it has been the policy of choice for both environmental groups and industry leaders in the United States and other countries as well.

In 2010, the National Research Council, which is a branch of the National Academies, issued a report entitled "Limiting the Magnitude of Climate Change". This report noted that "Meeting internationally discussed targets for limiting atmospheric greenhouse gas concentrations and associated increases in global average temperatures will require a major departure from business as usual in how the world uses and produces energy". Among the report's conclusions was the finding that the most effective policy strategy for the United States to pursue would be a strong, economy-wide carbon pricing system in the form of cap-and-trade, taxes, or some combination of both. The authors also recommended a portfolio of complementary policies aimed at ensuring rapid progress in the areas of: (a) promoting widespread implementation of existing technologies for energy efficiency and low-carbon energy sources (such as renewables); (b) advancing full-scale demonstration of carbon capture and storage and new-generation nuclear power, to determine whether these are truly feasible as major parts of our nation's future energy system; and (c) accelerating the retirement and replacement of existing GHG-intensive equipment and infrastructure.

Finally, which is the market system of choice, given the reality of the decision-making arena in which such policies are debated and implemented? The free market is indeed "free", which is to say, it is working all by itself, but it is not getting the job done. While the carbon tax is easier and less costly to design, implement and enforce than cap and trade, optimal taxes may have to be too high to be a politically feasible option at this time in the United States. Despite a multitude of problems, cap and trade programs such as the Environmental Protection Agency acid rain program have been implemented with some success in the United States.

6.6. Conclusion

Considering the traditional economic factors in isolation, the world is now in the beginning stages of a very important transition from artificially "low-cost" carbon-based fossil fuels, upon which it has been relying for many decades, to higher-cost alternatives, specifically to alternative fuels that do not lead to emissions causing climate-change. This transition is likely to involve substantial challenges, including energy shortages, increase in the cost of various fuels and a worldwide search for new energy sources.

The outcome of the transition from oil, coal and natural gas to alternatives will have profound impacts on every fuel-dependent national economy as well as on individuals, industry, governments and a wide range of organizations. If we recognize that these low costs were artificial and that the previous actual costs — political, military, human lives and rights — were very high, then this transition can be placed in a more correct context that more clearly indicates that it will benefit human kind not only in terms of climate change, but also in terms of political and human rights. The days of cheap energy that have fueled economic growth worldwide for many decades will soon be over, and the transition to new energy sources will be costly and challenging. Economic factors have played an important role in dealing with energy and climate change issues and they can play an important future role in addressing both these issues and in putting them in the broader proper context where the increasing energy costs are offset by the greatly decreasing political, military and human costs associated with supporting regimes that traditionally have provided the "low-cost" energy. Thus, total economic policy and factors such as allocations of funding between energy research and political regimes and wars in various regions should be included in addition to economic resources of prices and markets, along with the reduced effect on global warming that will result from finally giving up our long standing dependence on fossil fuels.

CHAPTER 7

WHERE DO WE GO FROM HERE?

Earth's orbit around the Sun — or for that matter any other solar orbit — is determined by a precise balance between the tendency of a body in space to fly off in a straight line and the gravitational force of the Sun, which continually bends the line into an orbit. One result of that balance is that the amount of time it takes to make one complete circuit — that is, one year in the life of a planet — depends entirely on the planet's distance from the Sun. The closer to the Sun, the faster a body must move to stay in orbit and the shorter its year. If a body between the Sun and the Earth tried to orbit the Sun in an elliptical orbit in one Earth year, it would be moving too slowly for its orbital speed to balance the Sun's gravity and would spiral outward into a larger ellipse.

However, there is an exception to this rule. There is one point on a straight line from the center of the Earth to the center of the Sun where a body could orbit the Sun in 365 and 1/4 Earth days. That point in space, known as the Lagrangian point L1 (named after the French mathematician Joseph-Louis Lagrange), is about one hundredth of the distance from the Earth to the Sun. At that point, the outward pull of the Earth's gravity just balances the Sun's gravity if the body has the right speed.

In fact, one of the technological fixes that has been proposed to alleviate further global warming is to place a giant parasol 2,000 kilometers across — big enough to block a few percent of the Sun's radiation — at L1. From the Earth it would look like a permanent spot on the face of the Sun and once in place it would free us to burn up all the fossil fuel on the planet without worrying about our climate.

This grandiose scheme, an example of what is known as "planetary engineering", is clever, amusing and probably possible. But it is a foolish idea. For one thing, Earth's climate is not an inevitable consequence of the flux of solar radiation at the position of our orbit. With the same solar flux, the Earth could be a ball of ice or a Venusian inferno. Changing the flux a bit cannot guarantee our safety. Of course, a parasol at L1 is just about the last thing a future energy starved world will need.

Shading the Earth with a giant parasol may not be a great idea, but if we are going to burn up all the fossil fuels we can lay our hands on, limiting the damage caused by carbon dioxide (CO_2) is an excellent one. How can that be done?

Carbon is the essential element of life. Every molecule of every bit of living matter contains carbon. Life is nourished by photosynthesis, the process by which plants consume sunlight, water and CO_2 and produce the stuff of which they are made. Thus, growing plant life removes CO_2 from the atmosphere. Unfortunately, it does not remove it for long; CO_2 is returned to the atmosphere when the plants rot or burn. Where else does CO_2 come from? From fossil fuels of course. One ambitious idea is to remove the CO_2 from the exhaust gases of fossil fuel burning power plants and sequester it somewhere. But where would that be?

One suggestion is to use the CO_2 to pressurize oil and natural gas wells. The pressure would help to push the oil and gas out, allowing the CO_2 to stay behind in the well. There is nothing wrong with that and, in fact, oil companies are already doing it to a limited extent, but oil wells do not offer the necessary storage capacity. We have already increased the concentration of CO_2 in the atmosphere by about hundred parts per million: If we had some way of extracting the CO_2 from the atmosphere and storing it, we would need to quarantine about a ten-thousandth of the atmosphere of the entire planet just to return the atmosphere to its preindustrial condition. That is a lot of gas to park somewhere. Even if we just try to stabilize the atmosphere, we will have to put away several times that amount before the planet's fossil fuel starts to be depleted.

The only place that would have the necessary carrying capacity is the ocean floor. In the atmosphere, CO_2 is a gas, but when subjected to the pressure and temperature of the deep oceans it would become a liquid denser than water and would sink to the bottom. Compared with the gaseous form, liquid CO_2 is very compact. It would take up a volume hundreds of times smaller than the equivalent gas, and the ocean floors cover most of the globe. There would be enough room.

However, the scheme of sequestering CO_2 and injecting it into the deep oceans might have some drawbacks. There are all sorts of life in the deep oceans and a layer of liquid CO_2 might snuff it out. Moreover, some of the

CO_2 will dissolve in the seawater and gradually diffuse upward. An increased concentration of CO_2, will alter the acidity of the oceans somewhat, which could be a very big headache for a lot of marine life, large and small. Also, any CO_2 that found its way to hot spots and mid-ocean ridges, where volcanic magma bubbles up, would be vaporized and might make it to the surface and escape. Finally, although the liquid CO_2 would be physically stable on the cold parts of the ocean floor, it is even more stable as a gas in the atmosphere. The liquid at the bottom of the sea is in a metastable state. Any event that stirs the oceans waters — say an asteroid impact — would release a large quantity of CO_2 to the atmosphere, with possibly catastrophic results. The CO_2 drifting over the land would suffocate all the life forms it encounters, including us.

The Kyoto protocol, signed by the U.S. in 1997, was a first attempt to do something about fossil fuel induced global warming. It was not very strong and even then the U.S. withdrew from it in 2001 because it threatened our economic wellbeing. There have been a number of other attempts along those lines since but none of them really sufficient to halt the warming trend. The world still awaits an agreement strong enough to do the job. What other possibilities are there?

7.1. Nuclear Power

As we reach the end of the age of fossil fuels, the world will have to consider increasing the use of nuclear power. It may be the only proven technology capable of filling in for the loss of those fuels on a massive scale. As we were reminded by the devastating events at Japan's Fukushima Daiichi nuclear power complex following a severe earthquake and tsunami, there are difficulties and dangers associated with nuclear power, but there currently may be no alternative.

The atomic nucleus is a tiny speck at the heart of every atom, which contains nearly all of the atom's mass. It contains protons, which have positive charge and, therefore, hate to be at close quarters with one another and neutrons, electrically neutral particles that tend to keep the unhappy protons together. The position of an atom on the periodic table is entirely determined by the number of protons in its nucleus. However, the number of neutrons in the nucleus of a given element is not so rigidly fixed. All elements have atoms with different numbers of neutrons — and, therefore, different masses. These are called "isotopes".

For example, the element uranium has 92 protons in its nucleus. Its most commonly found isotope has 146 neutrons, for a total atomic mass of 238 — that is, it has 238 times the mass of a single proton (protons and neutrons have nearly the same mass). We will use the scientific designation ^{238}U for that isotope. Another isotope, ^{235}U (with three fewer neutrons) comprises approximately 0.7% of natural uranium. This rare isotope has the important property that if it captures a slowly moving free neutron, it breaks violently apart, usually into two large nuclear fragments, plus, on the average, 2.43 free neutrons.

These fragments have lots of energy owing to a small mass defect in the reaction. A large part of the energy goes into the kinetic energy of those fragments, but the neutrons are pretty fast too. If the neutrons can be slowed down before they escape, they can be captured by other ^{235}U nuclei, producing more free neutrons, which in turn can collide with yet other ^{235}U nuclei and so on. That is what is called a "chain reaction". It will occur explosively, if the uranium is nearly pure ^{235}U.

The best way to slow down the neutrons and thus initiate the chain reaction is to make them collide with something not too much heavier than themselves. If they collide with a big, heavy nucleus they bounce off like a rubber ball recoiling from a concrete wall, keeping nearly all of their energy. But if they bounce off something lighter, that object is set in motion and removes some of their energy. In most nuclear reactors, water is used for that purpose. The water also serves to cool the fuel, which is heated because of the energetic nuclear fragments. That heat can be used to drive a turbine. Thus a nuclear reactor is just a source of heat, like burning coal.

The reactor that melted down at Chernobyl had a cooling system based on graphite, which is more commonly found in reactors used to produce weapons, not electric power. The six reactors that made up Japan's Fukushima Daiichi complex used water for cooling. Water-based cooling systems are generally considered safer than those using graphite, and the damage to Fukushima, while extensive, did not rise to the terrifying level of Chernobyl. Investigations into what happened at Fukushima — and the relative contributions of the magnitude nine earthquake, the subsequent tsunami, inadequate industrial oversight and aging infrastructure, among other factors — are continuing. Their outcome will undoubtedly shape ongoing debates over the future of nuclear energy. Today in the United States, about half of our electricity is generated by burning coal, with nuclear power contributing about 20%. In other countries, the amount of nuclear power varies from most (in France, nuclear energy provides some 78% of its power) to none (Italy, which has outlawed nuclear power plants, but buys electricity from French nuclear plants). Suppose the world were prepared (or forced) to ignore all the difficulties associated with nuclear energy and forge ahead to replace all stationary fossil fuel power plants with nuclear power. That would certainly help relieve the CO_2 problem in the atmosphere. Nuclear power would not replace oil for use in transportation, but let us suppose that the electricity it produces can be used to charge batteries or make an alternative mobile fuel such as hydrogen. Is there enough uranium around for that to be a long-term solution?

Estimating uranium reserves is not nearly as well developed a science as estimating oil reserves. Just as in the case of oil reserves at an earlier time, uranium reserves will surely increase as a result of both further exploration and better technology. However, known reserves are estimated to be enough to supply all of Earth's energy needs — at the current rate of energy consumption — for only 10 to 25 years. That estimate ignores the growing world demand for power as well as the Hubbert's peak effect, which is just as valid for uranium as it is for oil.

But that estimate assumes that only ^{235}U will be used to produce nuclear power. The vast majority of the Earth's uranium is ^{235}U, which is not used at all in conventional power reactors. It is possible to design a reactor in such a way that while it is producing heat, it is also converting ^{238}U to the isotope 239 of the element plutonium (Pu). Like ^{235}U, ^{239}Pu is fissile, that is, if it captures a slow neutron it can explode apart, yielding energy and neutrons. That kind of reactor is called a "breeder reactor". Breeder reactors would increase the amount of energy available from uranium a hundredfold. Unfortunately, plutonium is a very nasty stuff. In particular, it is much more easily converted from peacetime pursuits into weapons than is uranium. For that reason, there are no commercial breeder reactors in the United States, but Russia, Japan and India have experimental breeder reactor programs. Whether these nations will continue to pursue breeder reactor technology in the wake of the Fukushima Daiichi disaster remains to be seen, although early indications are that China is forging ahead with its program. Making the world safe for breeder reactors is a tall order.

There is at least one other possibility. Isotope 232 of the element thorium can be bred into isotope 233 of uranium. ^{233}U is also fissile. Nuclear reactors using thorium as fuel are still rare, but thorium is thought to be an important nuclear resource for the future because it is about three times more abundant on Earth than uranium. The natural radioactive decay of thorium and uranium is what keeps the Earth's interior hot.

Even if there is enough nuclear fission fuel on Earth to last for a while, the scale of what is needed is staggering. The largest practical nuclear power plant would produce about one gigawatt (one billion watts) of power. Just to replace the 10 terawatts of fossil fuel the world burns today would require opening 10,000 new gigawatt plants, one a day for 30 years. And by then, of course, we would need to replace far more than 10 terawatts.

7.2. Nuclear Fusion

The most promising, and most elusive, solution to our energy problems is the controlled nuclear fusion. The fuel it uses would virtually last forever and would not contribute CO_2 to the atmosphere. The easiest reaction to create is the fusing of two isotopes of the lightest element, hydrogen, called deuterium and tritium. (Only hydrogen isotopes have names of their own; all other isotopes go strictly by the numbers.) The nucleus of ordinary hydrogen is a single proton. The nucleus of deuterium is one proton with one neutron attached. The tritium nucleus with one proton and two neutrons is never found on Earth, but it can be made; however, it is radioactive and lasts for only a few years before it decays spontaneously.

If a deuterium nucleus comes into contact with a tritium nucleus, the result is a helium nucleus (two protons, two neutrons) and an extra neutron. The extra neutron is fired off, releasing a great deal of energy. The problem is getting the nuclei of the two atoms together. Because they are both positively charged, they repel each other with tremendous force, which increases as they get closer. But if they can be contained in a gas that is sufficiently dense and hot, their random thermal motions can be fast enough to overcome the force of repulsion and they can be packed closely enough to have a good chance to fuse. We know it can happen as nuclear-fusion is what happens in the Sun.

In the Sun, the superheated gases that produce nuclear reactions are contained by the Sun's gravity. To fuse and create power on Earth, they must be contained in some other way, since no known material can withstand the extreme temperatures that fusion generates. One solution to the problem is a magnetic field, which acts like a somewhat leaky virtual bottle, preventing the superheated electrically charged particles from escaping, at least for a while. All this takes place in a vacuum chamber, keeping the hot stuff away from any material walls. So, the trick is to make a gas of deuterium and tritium in a magnetic field — a gas hot enough and dense enough and contained long enough to produce useful energy. But it takes a tremendous amount of energy to heat the gas to the point where that can happen and an enormous magnetic bottle to contain the gas long enough. Fusion reactions have been created this way on Earth, but never enough of them to produce as much energy as it took to make the fusion reaction happen in the first place. That so-called "breakeven point" has been the central goal of the field of fusion research for decades. Although remarkable progress has been made, the best that has been done on Earth so far is to get back about half the energy that went in.

Generating useable fusion power will require a huge facility. A collaboration between the United States, Canada, the European Union, Japan and Russia that might produce an engineering prototype of such a facility, called the International Thermonuclear Experimental Reactor (ITER) suffered a setback in 1998 when the United States withdrew because of the $10 billion price tag; however, it rejoined in 2003 and the ITER project is now set to be built in Cadarache, France and is scheduled to begin operations in 2018.

If magnetically contained nuclear fusion ever does become practical, the primary fuels will (at least at first) be deuterium and tritium. Tritium can be made in place from reactions of the outgoing neutrons with a lithium blanket around the reactor. Deuterium in abundant supply is found in seawater, and very likely we would never run out of it. Less is known about the supply of lithium, which is found in a number of ordinary minerals.

Other approaches to containment are used by the National Ignition Facility at Lawrence Livermore National Laboratory in Livermore, California and the Z facility at Sandia National Laboratories Albuquerque, New Mexico. These United States government sponsored programs are attempting to produce short bursts of fusion energy by heating deuterium and tritium fuel pellets with intense laser pulses (LLNL) or electrical discharges (Sandia). Like magnetic containment, these so-called inertial containment schemes are big and expensive. They too have their optimistic supporters, who hope they will reach breakeven in the next decade. However, reaching breakeven would still leave us a long way from a practical power plant.

7.3. Solar Power

Beyond fossil fuel and nuclear power, all that remains is sunlight. We have, of course, always used sunlight — for example, to grow trees whose wood can be burned for warmth. One indirect solar source is hydroelectric power, a good example of seemingly renewable energy. The enormous pressure of water in a reservoir provides the force to drive a water turbine, which generates electricity. The Sun causes the water to evaporate, lifting it into the clouds, whence it can rain into the watershed and flow into the reservoir to produce more electricity. Early in the twentieth century, hydroelectric power seemed to be the way to go and dams to produce hydroelectricity were built wherever the appropriate conditions existed, including Hoover Dam on the Colorado River and the Grand Coulee Dam on the Columbia River. Today about 10% of United States power and about a quarter of the world's electric power is generated that way. However, we have virtually come to the end of our ability to increase the generation of hydropower. Dams have pretty much been built everywhere in the world they can be, so we cannot increase our production of hydropower enough to replace our dependence on fossil fuel. Furthermore, in recent years, we have come to appreciate some of the disadvantages of damming up rivers so that, worldwide, dams are probably being disassembled faster than they are being built. Finally, hydropower is not truly renewable in the sense of a resource that will last forever. All reservoirs eventually silt up. After a few hundred years even Colorado's mighty Hoover Dam will be nothing but a concrete waterfall.

Wind power is another indirect form of solar energy. As of 2015, just about 4% of the world's electric power was generated by the wind. That amount will increase because technological improvements and tax breaks for power producers using renewable sources have made wind power economically competitive with coal-fired plants. In Northern Europe, where there is plenty of wind and fossil fuels are expensive, wind may someday rival hydropower as a source of electricity. In fact in Denmark, 40% of electric power derives from the wind. However, many people regard wind farms as ugly and undesirable, and there are a limited number of places in the world where the wind blows strongly and steadily enough to be useful. We may inherit the wind, but we will not be able to live on it.

Solar cells can convert sunlight to electricity directly. However, gathering solar energy at the Earth's surface is a little like gleaning wheat from already harvested fields. The flux of light that reaches the surface of the planet is relatively weak and intermittent at best. Above the atmosphere, it is about eight times what it is, on the average, at the surface. The loss of intensity is partly due to reflection and absorption of energy by clouds and the atmosphere, but mostly it is because the light gets spread out over the curved surface of the spinning Earth. This observation has led to a number of schemes that would intercept the Sun's light in space and transmit the energy to the Earth.

One idea, studied in the 1970s under the sponsorship of NASA and the United States Department of Energy, was to put an array of solar cells the size of Manhattan into a so-called geosynchronous orbit — that is, an orbit high enough so that the array would remain fixed above a spot on Earth. It would, of course always face the Sun, and it would be positioned so as to stay out of Earth's shadow. Its electrical energy would be transmitted to the Earth in the form of microwaves, a part of the electromagnetic spectrum used for radar because of its cloud-piercing properties. Proponents estimate that more than half the electric energy generated at the solar array could be converted into electric energy in a receiving station on Earth roughly eight miles by six miles in area. About 800 such satellites in orbit, each with a receiving station on Earth, would be needed to provide as much energy as we use on Earth today. Other schemes of this sort have been studied more recently, using more advanced technology. It is conceivable that space-based solar power may someday make some contribution to world energy resources.

7.4. Improving What We Have Got

The best, most conservative bet for ameliorating the coming fuel crisis is the gradual improvement of existing technologies. To take just one modest example, at the end of the nineteenth century, Thomas Edison's invention of the incandescent light bulb became the very symbol of the dawning age of electricity. The incandescent light bulb works by passing an electric current through a thin resistive filament, heating it to a temperature so high that it radiates white light, like the face of the Sun. That is to say, the principal product of a light bulb is not light, but heat. Producing heat is the most wasteful possible way of using electricity. Only 1% or 2% of the electric energy consumed by an incandescent bulb turns into visible light.

A variety of more efficient light sources have been invented since Edison's day, although incandescent bulbs still rule the night. One new type — the light emitting diode (LED) — is worth a careful look. In certain kinds of materials called semiconductors (the stuff of transistors) an electric current does not just generate heat; instead it causes quantum mechanical events in which individual electrons absorb a well-defined quantity of energy from the current. When the electrons fall back into their original states, they give off, in the form of photons, all the energy they absorbed. Photons are light and depending on the particular properties of the semiconducting material, they can be visible light of various colors.

A few years ago, LED's were so inefficient that their only application was in alphanumeric displays like the flashing "12:00" that we all remember flashing on the face of our old VCRs because we never could be bothered to learn how to program the clock. The problem was not in the initial *electric current — excited electron — visible photon* reaction, which works very well, but rather in getting the photon out of the material before it bounced around many times and turned into heat. But clever engineers have solved that problem as a quick drive around many city streets and suburban neighborhoods will attest. In the last decade, the old-fashioned incandescent traffic lights in many municipalities have gradually given way to much brighter ones that have a kind of speckled look. These are arrays of LED's, which are so efficient that the savings in electric power easily makes up for the cost of buying the expensive devices.

Solar cells — also called photovoltaic devices or PVs — are just LED's running backwards. In an LED, you put electric current into a semiconducting device and light comes out. In a PV, you put light into the same kind of semiconductor, and electric current comes out. That would suggest that PVs could be made very efficient, turning all the light that falls on them into electric current, but that is not the case. The reason is that LEDs are efficient only at a single color — red, green or amber traffic light — depending on the particular semiconducting material they are made of. A PV, to be efficient, must be able to turn all of the light falling on it, of all possible colors, into electric energy. PVs that can do that do exist but they are very expensive and are currently used only in spaceflight applications.

The lesson to be learned from this story is that the most exotic objects (LEDs) can become commonplace (traffic lights) almost without our noticing it and they can suggest the direction of future developments (cheap, efficient PVs). Such a development could, for example, make the Saudi Arabian desert more valuable for the sunlight falling on it than for the oil buried beneath it. However, the scale of what is needed is breathtaking. Using present-day PV technology, in order to replace all the power generated from fossil fuels, an array spread over more than 200,000 square kilometers would be needed. That is an area roughly half the size of the state of California. All of the PV's made up to now would probably cover fewer than 100 square kilometers.

In 1986, the scientific world was astonished to learn of the unexpected discovery of high temperature superconductivity. An earlier scientific world, around 1911, had been even more astonished to learn of the discovery of low temperature superconductivity. It turned out that at temperatures a few degrees above absolute zero, many metals quite suddenly become capable of conducting electricity with no resistance at all. The 1986 discovery was that certain complex materials were capable of performing the same trick at a much higher temperature (albeit one still very far below ambient temperature anywhere on Earth).

The discovery of high temperature superconductivity immediately evoked pictures of a worldwide electric energy grid, which would serve to balance demand for electric power between day and night. As we noted in Chapter 3, electric energy is hard to store in large quantities and so must be generated on demand, which is much higher during the day than at night. Since it is always day on one part of the planet and night on another, a worldwide electric grid would solve that problem. However, electric power lines using high temperature superconductors have not yet proved feasible. They are subject to the same drawbacks as were earlier plans to use low-temperature superconductors for power transmission (the need for elaborate refrigeration and insulation, back-up systems in case of catastrophic failure, and so on). To make matters worse, high temperature superconductors are composed of more exotic, more expensive materials than the old, low temperature ones. Still, the astonishing discovery of high temperature superconductivity serves to remind us that nature may still have surprises for us that can change the landscape.

As this brief survey suggests, there is no single magic bullet that will solve all our energy problems. There is no existing technology, short of nuclear fusion, capable of replacing the oil we will eventually be without, nor is there any on the horizon that we can depend on to replace the fossil fuels when they are exhausted. If we permit them to become exhausted before we replace them, we may place the climate of our planet in grave danger. The best hope for our civilization lies in technologies that have not yet arisen — possibly based on scientific discoveries that have not yet been made. Most likely, progress will lie in incremental advances on many simultaneous fronts, based on principles we already understand: controlled nuclear fusion, safe breeder reactors, better materials for manipulating electricity, more efficient fuel cells, better means of generating hydrogen and so on. Developing those technologies will require a massive, focused commitment to scientific and technological research. That is a commitment we have not yet made. We urgently need to make it.

7.5. Some Further Thoughts

Four and a half billion years ago, chunks of rock coalesced out of the primordial gas and dust cloud circling the Sun and clumped together to form the Earth, a spinning sphere in a nearly circular orbit 93 million miles from the Sun. Those facts alone did not ordain our planet's destiny. It could have become nothing more than an icy wasteland, reflecting back into empty space most of the solar light that falls on it — and indeed, at some periods in its deep past it may have been just that. Or, it could have turned into a poisonous, lifeless inferno, like its near twin, Venus. Instead it became a balmy garden planet, with an oxygen-rich atmosphere and much of its carbon sequestered in the ground in the form of coal and other fossil fuels. It was life itself that helped to create this planet wide garden of Eden.

Life on Earth is sustained by a great underlying drama of radiant energy from the Sun — radiant energy that drives vast currents in the atmosphere and the oceans, warms the soil and in a thousand other ways undergo its inexorable entropic transition into ambient temperature heat, eventually to be radiated back out into space as invisible infrared rays. But that exchange of radiant energy happens on Venus, Mars and every other orbiting body in our solar system without nourishing life along the way. Now, in the last few instants of those four and a half billion years, life has turned intelligent enough to dig up and burn those fossil fuels. Humans will not be banished from the Garden of Eden, but we may very well end by destroying it.

In the meantime we have what may be an even worse problem. Our way of life, firmly rooted in the myth of an endless supply of cheap oil, will eventually come to an end when the final Hubbert's peak appears. But is it possible that this view is mistaken? Are there not experts on the other side who believe that the end of the age of oil is still far away?

Yes, there are. The oil companies, of course, study this problem, since their fate depends on it. BP is one of the world's more forward-looking energy companies and it refers to itself not as "British Petroleum" but rather as "Beyond Petroleum". It publishes a website full of useful information, not only on oil, but also on other fossil fuels, nuclear energy and renewable sources of energy. According to that site, the ratio of known reserves of oil to rate of use — the R/P (Reserves to Production) ratio — is 40 years. The R/P ratio for natural gas is 60 years.

The United States Department of Energy gives the R/P ratio for oil as 96 years. United States does not have a department of entropy.

The crucial difference between the Hubbert's peak prediction and the BP prediction is not a matter of how much oil there is, or of how fast we are using it. They pretty much agree on those questions. The difference is in when the crisis will occur. The unstated assumption of the BP prediction is that we will be fine until the last drop is pumped out of the ground. The Hubbert's peak prediction says that once we reach the halfway point, consuming the easily found oil in the ground, existing oil fields will start to become exhausted faster than new oil fields can be tapped. The rate at which oil can be pumped out of the ground will start to decline, including extreme sources such as fracking. That is the essence of the bell-shaped curve hypothesis. The Hubbert's peak assumption is that the crisis will occur not when the last drop is pumped but at the half-way point, where falling supply meets increasing demand. When we have consumed half the oil that ever was, that will be the time of crisis.

There is no doubt at all that the essence of the Hubbert's peak prediction is correct. It is possible, of course that the quantitative predictions are off so that the crisis will not occur until the next century or the one after that. The difference might seem important to us, but in the long view of history, a difference of 100 or 200 years means nothing at all. We, or our children, or our grandchildren will face some very difficult times.

If the problem were widely understood and acknowledged, we could go a long way toward easing the pain that the crisis will cause. We Americans are profligate users of energy. There are many ways in which we could reduce our consumption of fuel without abandoning our comfortable way of life. That would give us more time to convert to a temporary methane-based technology while we build up our capacity for tapping other fuel supplies.

In the long run, even those steps will not be enough. The real challenge — the challenge we would set ourselves if we had been courageous, visionary leadership — would be to kick the fossil fuel habit as soon as possible. In 1960, John F. Kennedy challenged us to put a man on the moon in that decade. And we did it! That was possible because we already knew the basic principles of how it could be done. There were formidable technological obstacles to overcome, but we are very, very good at overcoming that kind of obstacle when we put our mind to it. The energy problem is of exactly that nature. As we pointed out in Chapter 4, that is precisely what we need now.

We can envision a future in which we live entirely on nuclear energy and solar energy as it arrives from the Sun. That would not require a reversion to an eighteenth century lifestyle and a concomitant drastic reduction of the human population of the Earth. Instead it would be based on a sophisticated technology that converts sunlight and nuclear energy efficiently into electricity for stationary uses and produces hydrogen fuel or charges advanced batteries for mobile uses. That would leave the carbon in the ground or at least unburned, as a source for the petrochemicals that are also an indispensable feature of our way of life. It might be no more difficult to accomplish than putting a man on the moon.

Unfortunately, our present national and international leadership is reluctant even to acknowledge that there is a problem. The crisis will occur and it will be painful. The best we can realistically hope for is that when it happens it will serve as a wake-up call and will not so badly undermine our strength that we will be unable to take the giant steps that are needed.

Kenneth Deffeyes, a leading academic geologist, well schooled in the realities of oil and gas, realized that geophysicist M. King Hubbert had been right — that is, a peak in oil production for the Lower 48 had been reached — when he read a brief sentence in the *San Francisco Chronicle* in the spring of 1971: "The Texas Railroad Commission announced a 100% allowable for next month". Such a quiet pronouncement would have slipped by most readers, but to an insider like Deffeyes the words were momentous. The quaintly named Texas Railroad Commission, after all, was the cartel that controlled production in the United States oil industry by manipulating the excess pumping capacity of Texas wells. By announcing a "100% allowable" the commission signaled that Texas no longer had any excess pumping capacity. The Texas oil fields could be pumped flat out because they had reached the point of diminishing return — and consequently the Texas Railroad Commission had lost control of the market.

After that the world oil market fell under the domination of another cartel, the 12-nation Organization of Petroleum Exporting Countries (OPEC), indeed modeled after the Texas Railroad Commission and led by Saudi Arabia. The Saudis have been manipulating the price of oil ever since by strategic use of their excess pumping capacity. Thus, the news to look for, signaling the arrival of a *worldwide* Hubbert's peak would be that Saudi Arabia no longer had any excess capacity.

Exactly that news appeared on the front page of the *New York Times* on February 24, 2004. Headlined "Forecast of Rising Oil Demand Challenges Tired Saudi Fields" and written by Jeff Gerth, the story went on to say "The country's oil fields are now in decline, prompting industry and government officials to raise serious questions about whether the kingdom will be able to satisfy the world's thirst for oil in coming years". Like many *New York Times* articles, this one was very long and regularly contradicted itself in an apparent effort to achieve what is known as "balanced reporting". So, much further along, Gerth wrote, "Some economists are optimistic that if oil prices rise high enough, advanced recovery techniques will be applied, averting supply problems". In the very next paragraph Gerth takes it back: "But privately, some Saudi oil officials are less sanguine". And so on. We now know that advanced recovery techniques have indeed extended the world's supply of oil, but no one knows for how long.

OPEC's policy over recent decades has been to control oil supply so as to keep the price of oil not just above some chosen minimum, but also within a certain range — not too low but also not too high. The reason for trying to keep the price from going too high is partly not to discourage demand for oil, but also to prevent investment in alternative fuels. The implied threat is that if you invest money to develop a competitor to oil, we will flood the market with cheap oil and wipe out your investment. However, if the Saudi fields have really peaked, that becomes an empty threat, and the cartel stands to lose control of the market.

The United States is not a member of OPEC, although it played a role in its establishment in Baghdad in 1960, but the United States government shares OPEC's goal of keeping an upper limit on the price of oil; voters tend to get very unhappy when the price of gasoline at the pump rises. If Saudi Arabia can no longer flood the market, where would the extra oil supply be found? Canada (also not an OPEC member) now claims the world's second largest reserves after Saudi Arabia, but those are largely locked up in oil sands, solid deposits that must be mined not pumped and so will not be flooding anything anytime soon. The world's third largest reserves are claimed by Iraq, where a 100 billion barrels of oil are waiting to be exploited. Under the regime of Saddam Hussein, however the spigot was broken. Although few seem willing to talk about it, that was certainly one of the big reasons for the United States invasion of Iraq in 2003: The idea was not so much to steal the oil from the Iraqi people — they will be allowed their small profit from the raw material before the real money gets made. Rather, the idea was simply to get Iraqi oil back on the market.

Many experts doubt that there will be an oil crisis in the near future. They have been dubbed the "antidepletionists". In our experience, they are intelligent, well-informed people — and most of them are employed by or have ties to the oil industry. That does not automatically make them wrong. After all, people who work for the oil industry are the ones most likely to be interested and knowledgeable about it. We should keep in mind, though, that the oil industry has a very strong incentive to deny any looming shortage of oil. The reason is to keep the price down on properties they would like to acquire.

As we have seen, the worldwide "proven reserves" of oil now stand at just over one trillion barrels, and the R/P (reserves to production) ratio is about 40 years. There is nothing alarming about that, say the antidepletionists; the R/P ratio hovered at around 40 years through most of the twentieth century. That is true, but to understand what it really means we have to reexamine the term "proven reserves". To most of us, "proven reserves" would consist of all the oil that has been discovered minus all the oil that has already been extracted. But that is not how the oil industry uses the term. Oil companies and petroleum producing nations alike report as "proven reserves" only a portion of what they believe themselves to have in reserve. When a new field is discovered, geologists use various techniques to measure its length and width, its depth, the porosity of the rock and so on, finally coming up with an estimate of how much oil the field may contain. That estimate gets turned over to the officials of the company or country, who can report as "proven" whatever fits their current needs, saving the rest for a rainy day. That leeway is what permits "proven reserves" to keep on growing and the R/P ratio to remain essentially constant no matter what is happening in real oil fields.

What is actually going on in real oil fields is sobering. Worldwide, the rate of discovery of conventional oil peaked around 1960 and has been declining ever since. As we have seen, unconventional sources have boosted the total available. Meanwhile, the worldwide rate of consumption of oil has continued to grow. In the meantime reserves have steadily increased, partly because of new techniques, but also because companies and countries continue to pull out new reserves they have kept up their sleeves. In fact, in the late 1980s the proven reserves of OPEC nations jumped by nearly 400 billion barrels without the benefit of *any* new discoveries! To reach that new height, OPEC merely changed its quota rules for how much each nation was permitted to pump based in part on their reported proven reserves, and the new proven reserves magically appeared.

Something like 500 billion barrels have been brought out of the shadows by these methods and added to worldwide proven reserves over the past 30 years — an amount equal to half of all existing reserves. Obviously, this game cannot go on much longer. Either the industry will expand its unconventional recovery or it will simply start lying — reporting reserves that simply do not exist. That has apparently already started to happen. The once proud Royal Dutch Shell Group made headlines a few years ago when it was forced by outside auditors to reduce its claims of proven reserves — and correspondingly, the value of its stock shares. Companies do use outside auditors, but needless to say, countries do not.

Some people are fond of saying that discovery had been declining since 1960 because so much oil had already been found that no more was needed; thus exploration dwindled to a standstill. That is most certainly not the case. For example, 1999 and 2000 were spectacular years for oil discovery, driven by giant findings at Azedegan in Iran and the Kashagan East field in the North Caspian Sea. But even in those years, new discovery fell short of consumption. In truth, the world is consuming oil at such a breathtaking rate — more than 25 billion barrels per year and rising rapidly — that no discoveries, past, present or future, are going to keep up with demand. As we noted in previous chapters, the era of private auto ownership is just beginning to heat up in China and India.

Economists believe that the demand for anything can never exceed its supply. The mechanism of price assures that the supply will always show up when it is needed. Of course that is pretty much never been true of the oil industry, which has nearly always been governed by cartels, first the Texas Railroad Commission, and now OPEC. When world oil production peaks, OPEC will lose control and the price mechanism will kick in with a vengeance, making it economically feasible for other sources of fuel to replace the missing oil, assuming that other sources of fuel exist.

In a sense that is already happened in the case of Canadian oil sands, which were being mined at a profit when the price of oil was high. What came out of the ground was solid ore, but the product that comes out of the ore is not rich enough to make gasoline, so hydrogen must be added. As a result, some of the world's largest plants for extracting hydrogen from natural gas have been built in the province of Alberta. In other words, oil from oil sands is not only more costly in money than conventional oil; it is also more costly in energy. That will be increasingly true as other hydrocarbon resources are exploited.

Thus, if we are willing to let the planet's climate fend for itself while we go on merrily burning fossil fuels at ever increasing rates, and if we are willing to pay ever-higher prices in both cash and energy, we may be able to muddle through for much of this century — that is, provided global political and social stability can somehow be maintained in the face of the huge fuel costs and the economic dislocations that will occur.

Meanwhile, in the current (possibly) pre-peak period, our enormous consumption of conventional oil makes us wholly dependent on some pretty dicey places in the Middle East. Perhaps truly at issue here is not the debate between those who think there's plenty of oil out there and those who do not, but a quite different kind of question: Which comes first, Hubbert's peak or the collapse of the Saudi regime? Both would have the same effect and both seem inevitable.

To sum up, there are three good reasons for trying to kick the fossil fuel habit as soon as possible: First, our present dependence on cheap oil makes us subject to events in some very unstable parts of the world; second, burning up all the fossil fuels we can get our hands on could cause irreversible damage to the climate of the only planet we have; and third, since the stuff will eventually run out in any case, we should give ourselves the best head start we can in preserving what is worth preserving in our civilization. Kicking the fossil fuel habit will require harnessing and organizing the creativity and ingenuity of scientists, engineers, social scientists — indeed, all of us — all over the world.

This is no small task. We do understand the underlying scientific principles that will solve the problems, but we do not know what kind of solutions will prove both technically possible and socially feasible. We cannot let the choice of solutions be dictated by some central authority; neither can we leave the solution up to the marketplace. If future generations are to thrive, we, who have consumed the Earth's legacy of cheap oil must now provide for a world without it.

CHAPTER 8

THE FUTURE OF ENERGY BASED
ON TECHNOLOGICAL INNOVATION

8.1. Why We Need a New Manhattan Project to Address Energy Use and Climate Change

Albert Einstein once observed that, "Everything has changed since the day that the power of the atom was unleashed except for one thing: The way we think". We believe that this observation also applies to the energy/climate change problem as addressed starting with the Kyoto agreement and on up to the Paris accords of 2016. We seem to be mired in the approach of attempting to directly reduce emissions of carbon dioxide (CO_2) and other greenhouse gases through international agreements while not looking at other possible ways of dealing with this global challenge. The 2009 Copenhagen Conference accords, whose centerpiece was a non-binding agreement spearheaded by the United States, to limit global increase in temperature, would, in our view, simply repeat the history of the Kyoto Protocol. The conference was highlighted by many speeches and inspirational pledges, but it reached no definitive consensus on how to bring about real change to a situation that we believe will only get worse. The later Paris accords have not yet been endorsed by enough countries to bring them into effect.

The major CO_2 emitters today include countries with sizable populations and relatively low CO_2 per capita ratios, especially China and increasingly India and other large developing economies such as Indonesia, Pakistan, Brazil, South Africa, Nigeria and Bangladesh. Asking these nations to reduce their CO_2 output while the rich nations, including the United States, the member states of the European Union (EU) and other nations continue to produce greenhouse gases simply will not work.

We believe that we need a totally new approach to the problem — one that recognizes that we are facing a challenge that cannot be solved

79

by political or economic means alone. Our proposed alternative would be a major scientific study of CO_2 and other greenhouse gases in the atmosphere (methane, water vapor, nitrous oxide, ozone, etc.) that would develop a long-term strategy to allocate investment and craft the necessary technologies, policies and institutions to address this problem via a crash R&D program. Precedents for such a crash program include the Manhattan Project to build the United States atomic bomb and Project Apollo to put a man on the moon as well as similar crash programs in Britain to build radar and in the Soviet Union to build missiles.

We would initiate this new approach by bringing together an outstanding team of physical and social scientists and engineers. They would focus on setting the policy agenda and research priorities to address this issue in the same spirit that an international, interdisciplinary team of scientists, engineers and mathematicians developed the atomic bomb in the Manhattan Project during World War II. The impetus for this development could be the realization on the part of political and scientific leaders that the piecemeal plans proposed thus far to deal with these new realities relating to climate change simply will not work. Some areas for the team's work might include:

- Developing institutions for financing energy in a manner that enables governments to pool the risk and provide lower cost funding given that alternative energy projects are very capital intensive, with a relatively high cost of capital.
- A large and important barrier to developing alternative energy sources is the regulation of the industry at the national level. Although it is designed to reduce investor uncertainty and to shorten the time it takes to replace the coal burning generators that are producing more than 30% of America's CO_2 emissions, the current permitting process for both conventional and alternative energy sources makes adopting alternative energy sources prohibitively expensive. This significant barrier must be recognized and overcome.
- The cost of solar panels has dropped by almost a factor of four in recent years and it is expected to drop still more. The major component of the cost of solar power is now the balance between various energy system, in particular its integration into the current energy infrastructure and the infrastructure needed to transmit solar-generated power. In the United States, for example, more land is currently devoted to producing ethanol from corn than would be needed to fulfill the electricity needs of the

entire country if solar arrays and power plants were located in the Southwest. If we could lower the costs associated with establishing a balance among the various energy systems and fitting solar into the balance.

It has great potential to become a major CO_2-free source of electricity. We would emphasize, however, that technical barriers, including storage and transmission will need to be overcome, before extensive use of solar power can become a reality. The United States could follow the lead of Germany, Spain and Japan in their use of solar power all of which are ahead of the United States that ranks fourth in the world. The United States should follow the French example of adopting a standard blueprint for the design and construction of its future nuclear plants rather than designing a unique plant for each location. This proven approach would not only lower costs, but it would also simplify the regulatory process and make nuclear power more widely accepted, much as it is today in France. There are some projects that will be more difficult:

- The CO_2 bond requires enormous energy to break, as a possible new approach to directly eliminating excess CO_2, but nature has provided a wonderful mechanism to do so through photosynthesis, the process by which plants use sunlight to break CO_2 into its chemical compounds. There is already substantial research on using photosynthesis to make use of CO_2 to produce economically useful products. This research should be supported and encouraged. A cap-and-trade system that includes a market for sequestered CO_2 would make such projects economically viable.
- Developing transmission lines based on carbon fiber and developing efficient storage methods for electricity would greatly reduce our dependence on fossil fuels. CO_2-free base load power (where "base load power" refers to the minimum level of demand on an electrical supply system over 24 hours) is the foundation of a sound electrical system. Base load power sources are those plants that can generate dependable power to consistently meet demand. Such power could be transmitted both to meet peak loads and used to power vehicles.

This change in our thinking and the development of a new approach to the climate-change problem would be much more productive than the largely rhetorical promises of the Copenhagen Conference and similar approaches, which will likely only repeat the history of the Kyoto Protocol.

Returning to Einstein, he once defined insanity as doing the same thing over and over again and expecting different results. One wonders what he would have to say about the repetitive nature of all of our approaches, thus far, to energy issues and climate change. The Manhattan Project approach that we are suggesting would have the merits of distilling the best thinking and establishing priorities based on a comprehensive assessment, and then focusing both the public and the international scientific community on the results. It could be sponsored by governments or, perhaps even better, by a coalition of receptive and committed governments, non-governmental organizations, universities, foundations and others.

8.2. Other Policy Initiatives

These new approaches to energy production based on technological innovation can be usefully supplemented by other complementary policy initiatives, especially improvements in energy efficiency. The latter could include replacing every light fixture worldwide with highly efficient LED lighting. There are close to seven billion light fixtures in the United States alone, where lighting utilizes some 23% of all electrical consumption or about 230,000 megawatts of capacity. Assuming a 50% increase in lighting efficiency, these LEDs could reduce demand by 115,000 megawatts. This lighting change would cost approximately $650 billion, a retrofit that could be financed and paid back in a mere two to five years through the resulting energy savings, depending on the specific application and the local rates per kilowatt-hour.

In addition, a major program could be initiated for geothermal heating and cooling. In the United States, air conditioning represents 31% of electricity demand or 310,000 megawatts of capacity. Geothermal energy could increase the efficiency of such heating and cooling by 30% or 93 megawatts of savings. Solar-water heating and on-demand water heaters should also be implemented as has been done in nations such as Israel, where virtually every house has a solar water heater on its roof that is both efficient and reliable.

Yet another initiative would be to eliminate the investor-owned guaranteed rate of return utility model and replace it with the Independent System Operator (ISO) model. This ISO approach operates a wholesale power system that balances the need for higher transmission reliability with the need for lower costs. For example, the California Independent System Operator Corporation, located in Folsom, is a non-profit public benefit long-distance power lines that make up 80% of California's power grid and acts as a key platform to achieve California's clean energy goals. The ISO provides full market rates for residential and business producers of clean renewable solar, wind and biomass produced electricity. Permits should be issued by the ISO, however, only when a house or building has performed needed retrofits so the benefit cannot be achieved unless energy efficiency is fully addressed.

Another initiative would be to convert the current federal investment tax credit for alternative energy usage (including solar water and space heating, thermal electric and thermal process heat, as well as photovoltaics, wind, biomass, geothermal electric, fuel cells, geothermal heat pumps, etc. to a 30% investment tax credit for energy efficiency). It is far cheaper to develop energy efficiency than to invest in solar, wind and biomass, although, as noted earlier, we still strongly advocate programs to develop and deploy solar energy on a widespread scale. This initiative would accelerate the payback period on energy-efficient devices and substantially reduce the cost of subsequent alternative-energy installations. Again, the ISO model would help defray the cost because small producers, both households and firms, would be paid market rates for any excess capacity that they can sell to the grid. The cost of the energy-efficiency tax credit would be made up through the increased employment and the resulting payroll taxes from the retrofits noted above.

While electricity rates would rise initially under the ISO model, this approach would eventually induce additional energy savings through conservation because it incorporates time-of-day-pricing that prices electricity use higher at peak periods and lower at off-peak periods. Electricity rates would initially rise under the ISO model, but time-of-day-pricing incentives would ultimately encourage conservation by shifting demand to off-peak periods. Higher prices will also be offset through energy-efficiency retrofit savings. As more alternative energy capacity is brought onto the grid for sale, prices will fall because of supply and competition in the marketplace with the ISO approach. California's experience with ISO shows that utilities first turn on fixed cost assets, such as solar, wind, hydro and geothermal, followed by variable cost assets such as natural gas and coal gasification as demand expands through the day.

Small-scale solar, wind and biomass power are most efficient when they are installed where the bulk of the power will be consumed to avoid power losses in the course of grid transmission losses, which can be up to 10%. These losses necessarily reduce the load on the grid and the additional ratepayer/taxpayer expense for expanding the grid. A tariffic of 2% to 5% on alternative-energy excess production (i.e., non-usable energy) can fund grid maintenance and expansion where necessary.

The most important initiative that we would advocate is to stop burning coal that is now being used worldwide to produce electricity. As we have emphasized, there is no such thing as "clean coal", despite attempts by the industry to popularize this idea. A carbon tax would help enormously to wean such nations as China, the United States and the EU from their dependence on coal, but newer technologies, including solar with newly-developed thin-film solar panels, as well as refined systems for biomass and hydropower, will help contribute. All of these technologies could be part of a major initiative in technological innovation to reduce greenhouse gases that is long overdue.

SUMMARY

The availability of cheap, plentiful oil has established itself as the very root and symbol of our civilization. But that oil will eventually begin to run out. And with China, India, Brazil and other countries coming on line as automobile driving nations, the demand for oil will continue to grow exponentially even as the supply begins to diminish. In the meantime, the burning of fossil fuels among other things has dumped many tons of carbon dioxide, which is a greenhouse gas, into the atmosphere, leading to global warming, the loss of ice and glaciers in the far north and everywhere vast changes in our climate system.

The system of production and use of oil is a classic oligopoly, with a few suppliers and many, many users. That is true of the companies that extract oil, the countries that have oil and even of natural gas, an important substitute for oil. In the United States, attempts to break up the largest oil producing entities have met with mixed success.

Some 60 years ago, a geologist named Marion King Hubbert predicted that the United States would start to run out of oil around 1970. Hubbert's peak arrived right on schedule, but new, more perilous drilling techniques have deferred the inevitable. Nevertheless, other geologists started paying attention to Hubbert's ideas. Collectively, they have predicted that the world will start running out of oil sometime in the next century. Their predictions are subject to dispute, but they have made one very important point: Our real problems will arise, not when we have pumped the last drop of oil, but when we reach the half-way point and the available supply starts to diminish.

To understand the threats to international energy security, one must take into account the possibilities of oil disruptions, the issues that arise from growing dependency on natural gas, our relations with the Middle East (the principal source of oil) and the possibility of a major loss of supplies. Each of these issues deserve careful study.